FEEDING THE WORLD

Texas A&M AgriLife Research and Extension Service Series

Craig Nessler and Douglas L. Steele, General Editors

FEEDING
THE WORLD

Agricultural Research in the Twenty-First Century

Gale A. Buchanan

Texas A&M University Press   •   College Station

This paper meets the requirements of ANSI/NISO Z39.48–1992
(Permanence of Paper).

Binding materials have been chosen for durability.

Manufactured in the United States of America

LIBRARY OF CONGRESS CATALOGING-IN-PUBLICATION DATA

Buchanan, Gale A., 1937– author.

  Feeding the world: agricultural research in the twenty-first century / Gale A.
Buchanan. — First edition.

     pages cm. — (Texas A&M AgriLife Research and Extension Service series)
  Includes bibliographical references and index.
  ISBN 978-1-62349-369-1 (printed case: alk. paper) —
  ISBN 978-1-62349-370-7 (ebook)
1. Agriculture—Research—Finance. 2. Agriculture—Research—United States—
Finance. 3. National agricultural research systems—United States—History—
20th century. 4. Agriculture and state. I. Title. II. Series: Texas A&M AgriLife
Research and Extension Service series.
  S540.E25B83 2016
  338.1068′1—dc23

                         2015029142

General editors for this series are Craig Nessler, director of Texas A&M AgriLife Research, and Douglas L. Steele, director of the Texas A&M AgriLife Extension Service.

*This book is dedicated
to those individuals who
have engaged in any way in the
agricultural research process.*

"When tillage begins, other arts follow. The farmers, therefore, are the founders of human civilization."

—DANIEL WEBSTER

Contents

Author's Note

While this book is about agricultural research, the author wishes to make abundantly clear, the critical importance of the other components of the agricultural support system—agricultural cooperative extension and education-teaching programs. These three efforts comprise agriculture's blue ribbon support team. It is obvious that agricultural education is necessary for providing skilled workers for all aspects of agriculture including highly trained scientists to conduct agricultural research. Likewise, cooperative extension provides the vital linkage of research to the user. Without an effective cooperative extension service, the efforts of agricultural research would be marginalized. This is in keeping with Nobel laureate Norman Borlaug's admonition, "Take it to the farmer if you want impact." These three functions working in harmony are far more effective than the sum of each of the three functions working separately.

Preface

Agriculture is not just the most essential industry,
it is the only essential industry.
—BOYSIE F. DAY

Agricultural research has been a part of my entire life. I learned first-hand some of the problems and challenges associated with farming as I grew up on a small farm that produced an array of crops and livestock. The need for change, improvements, and the importance of new technology was apparent when I was a small boy. Controlling insects, diseases, and weeds as well as harvesting crops such as peanuts and tobacco without any mechanization validated this point. The importance of new technology became particularly evident when the first tractor arrived on our farm. While we kept the mules for a time, the tractor rapidly changed our approach to farming.

My father provided my first introduction to the concept of research because he was always trying a different rate of fertilizer, plant spacing or cultural practices. While these were quite simple "experiments" or demonstrations, they made an impression that there were other, perhaps better, ways of doing things.

After earning BS, MS, and PhD degrees, I was ready for my first assignment in research. I spent 15 years in weed science teaching and research. My research focus areas were control of weeds in crops, crop-weed interactions, and basic weed biology. The following 25 years I was engaged in administering various agricultural research programs including serving as an experiment station director in two states and as dean and administrative head of agricultural programs at one land-grant university. A final assignment in agricultural administration was as Chief Scientist, US Department of Agriculture, and Under Secretary for Research, Education, and Economics (2006–2009).

During my days as a researcher, I was oblivious to the financial challenge associated with research because I was fortunately successful in generating extramural funding for my personal research programs. However, once I became responsible for administering a wide range of agricultural research programs, I quickly realized that many areas of agricultural research could not be supported extramurally. Across the years I've seen the challenges in funding agricultural research grow severely. Particularly in recent years, the funding challenge has become critical—even dire. While this is serious, it is not my greatest concern. More importantly, there doesn't appear to be the sense of importance or sense of urgency about agricultural research that is commensurate with the challenges facing mankind or the planet.

My concern involves just about everybody. Farmers, one of the primary beneficiaries of agriculture research, extoll the importance and value of research but their greatest concern revolves around farm programs. I once had a leading peanut farmer say in an open meeting, "Dean, what you do in research is truly important, but the USDA commodity program is our life blood." The result has been that some farmers give only modest support for research through commodity support by "check off" programs.

Consumers, who are the true beneficiaries of agricultural research, for the most part, see no reason for even paying lip service to agricultural research. Consumers think as they do because there has always been an abundance of agricultural products available at a reasonable cost. Perhaps this is understandable since the final product of agricultural research is usually far removed from the original research. For example, rather than thinking of the innovative research that gave us wrinkle-free cotton fabric, recognition would probably go to the garment manufacturers. Credit for all the food on well-stocked shelves in the local grocery store goes not to the farmers, but to the wholesaler who supplies the particular grocery chain.

Perhaps the biggest disappointment is our political and governmental leaders. These are the people and institutions that should be providing the vision and basis for the future of our civilization. For example, I've often wondered why those members of Congress who have agricultural interest do not place a higher emphasis on agricultural research. I think it must be that farm programs and food assistance programs relate to a far greater part of their constituency. Of course, potential votes always take precedence. Other areas of research are simply closer to each member. Everyone has a medical problem or knows someone who does. Few can argue that it is more exciting to announce a possible cure for an important disease than increasing the yield of corn by one bu/acre.

Probably a similar argument could be made for various governmental offices. Perhaps because the agricultural research budget is so small, it does not register in science circles in Washington D.C. What is even more frustrating is the fact that a major agricultural research effort—determining the corn genome—was actually funded through the National Science Foundation (NSF) and not through US Department of Agriculture research channels.

The manner in which government is funded causes extreme challenges. The USDA receives a budgetary allocation and the Secretary of Agriculture must then develop the budget for all components of the department. Within USDA, the programs that benefit individuals, e.g. farmers through farm commodity programs, citizens through a multitude of nutrition programs, landowners through conservation programs and communities through rural development programs, obviously take a high priority. Regulatory programs are a must. Clearly, it is a "zero-sum game!" One area cannot receive money spent in another area. What this really means is agricultural research is seriously underfunded.

The crowning event that illustrated the magnitude of the challenge was the passage of the American Recovery and Reinvestment Act better known as the "Stimulus Bill" in 2009. Initial discussion of the legislation in Congress occurred while I was in the administration. On hearing that the bill provided for more than $800 billion to re-build our economy, I thought, "at last there will be better days ahead for agricultural research." I knew that in the mid-19th Century when our political leaders were developing plans to build a country and an economy they created legislation designed to strengthen agriculture and manufacturing.

In my naiveté, I could see mega-dollars going to engineering schools to study how we could make "widgets" as economically as China or Bangladesh with less labor. I could also envision major efforts to make agriculture more productive. This, of course, would require an enormous infusion of new information, knowledge, and technology which could be gained only through research.

This seemed so simple because to build an economy requires that we build something useful or for sale, grow something we can eat or sell or something that can be removed from the land such as coal, minerals, petroleum or harvested from the forest such as timber.

I don't know how much was spent in engineering schools to fos-ter manufacturing to build "widgets" more efficiently. However, I do know not a single penny was allocated to support agricultural research programs. Admittedly, there were some funds allocated for mainte-nance of government research facilities. We did spend mega-billions on roads, cities, bridges, and some areas of governmental research. The National Institutes of Health (NIH) received several billion dol-lars and the National Science Foundation (NSF) received more than $2 billion of one-time money for enhancing their research effort.

The current situation is of only modest concern if the world stays the same. Alas, this is not to be. We are already aware of changes looming on the horizon that are coalescing to form a different world in a few short years. Such changes include increasing population,

enhanced expectations by all people, energy requirements and un-
knowns regarding global climate change. Each of these has implications for agriculture and particularly agricultural research. I believe
we are facing the most important challenges to our civilization other
than perhaps some form of nuclear exchange. Everyone has a responsibility to meet these challenges. Consequently, the idea of this book
was born.

I'm very much like the turtle on the fence post. Whatever I have
done or accomplished, I owe much to others. First, my wife Carol,
for her steadfast support and for encouragement. Arden Bement, a
great friend and colleague, who at the time I served in the USDA
was Director of the National Science Foundation. He encouraged me
to put my frustrations about funding agricultural research in writing. Also, Ray Orbach, who was Under Secretary for Energy in the
Office of Science in the Department of Energy while I served in the
administration. He and I shared a common concern regarding the
importance of "finishing the research before implementing a technology." All my former fellow administrators, including experiment
station directors, deans and university administrators and, especially,
administrative heads of agriculture deserve special recognition. I'm
proud to have been a member of these groups.

I'm grateful for the University of Georgia, in general, and Dr. Craig
Kvien, professor, UGA Tifton Campus, in particular, for his support.
The exceptional typing skills of Mrs. Leanne Chafin were critical to
the writing of this book. Mrs. Evelyn Folds was exceedingly helpful in correspondence with my contributors and reviewers. I'm very
grateful for their support. I am also grateful for Eddie Gouge, Ian
Maw, The Association of Public and Land-Grant Universities, Sandy
Suit; Sara Mazie, Sandy Miller-Hays, Mike Arnold; U.S. Department
of Agriculture and Linda Drew, U.S. Department of Defense.

Also, Pat Jordan, former colleague and former Administrator
CSRS and Director of the Southern Region Research Center in New
Orleans, LA; Caird Rexroad, former Associate Administrator, USDA,

ARS; Ralph Otto, former Deputy Director for Food and Community Resources, National Institute for Food and Agriculture; A.G. Kawamura, former Secretary, California Department of Food and Agriculture; Stan Wilson, former Vice President for Agriculture at Auburn University who was my boss and always challenged me; Diane Rowland, Associate Professor, Crop Physiology, University of Florida; and Charles Stenholm, a great agricultural leader and former US Congressman from Texas, for reading an early draft of the book. Special thanks to Pat Jordan and Stan Wilson for reading a final draft of the book. I also greatly appreciate Roger Crickenberger, Gregory Weidemann, Louis Magnarelli, Melinda Geisler, Wendy Wintersteen, Kathryn Boor, Ford West, Keith Fuglie, Jack Payne, Rusty Griffin, Arden Bemet, Bob Herdt, Rakesh Singh, Neville Clarke, Scott Hutchinson, Ruben Moore, Jim Borel, Patrick Pickett and Jason Carver for reading selected sections or chapters.

I appreciate special help from Duncan McClusky, Cynthia Clark, Peggy Ozias-Akins, Jay Vroom, Jeff Pettis, Barb Glenn, Todd Peterson, Marc Rothenberg, Mary Bohman, Wally Huffman and Jim Hook for reading some parts of the book. I'm also indebted to many colleagues especially David Bridges and Jim Hook for advice, counsel and assistance in the preparation of a manuscript for publication.

I'm especially grateful for the contributions and reviews by Keith Fuglie, Economic Research Service for his help and guidance on the economic aspects of the book. Special recognition and thanks for Barry Jones, friend and former colleague for his professional editing.

Also, I must recognize and thank all scientists and staff, both state and federal for providing much of the content of this book.

While I'm grateful for all contributions and reviews, I had the last review and consequently any errors are mine alone for which I take full and complete responsibility.

FEEDING THE WORLD

Introduction

Research is to see
What everybody else sees –
But to think what
Nobody else thinks.
—ALBERT SZENT GYORGYI

Feeding the World: Agricultural Research in the Twenty-First Century is a bold and proper title for this book. This title was chosen because agricultural research is, without question, the most critical aspect of meeting the challenge of feeding the people of the world. As long as humans are dependent on food, fiber, and energy, agricultural research will remain essential and critical for the survival of *Homo sapiens*. This, in no way, diminishes the relevance and importance of all other research endeavors.

As someone trained in the biological sciences, I recognize the importance of and need for research. Of course, research relevance in the social sciences will continue to grow in importance. As we compare various research endeavors in the agricultural sciences, the thesis becomes clear and convincing: agricultural research forms the foundation for other avenues of inquiry; without such research our basic needs would not be met.

Medical research is important. Many of us have suffered a serious disease or know someone close who has. We know someone who has a serious debilitating ailment such as cancer, Alzheimer's, diabetes, heart disease, HIV or a multitude of other serious ailments. Everyone hopes for a cure. When such ailments become personal we desire a faster-paced research effort. Consequently, I do not diminish the importance of medical research. The fact is medical research saves lives, prolongs life, and contributes to a better quality of life. However, when you compare saving individual lives to saving civilization, it's an easy choice.

Defense research is the largest research enterprise supported by the US government today. First, it must be recognized that much defense-supported research has immediate and relevant applications that provide security and benefit for society. One of the greatest concerns in the world today is some form of nuclear exchange or rogue group detonating a nuclear device near a population center. This is not so much a researchable problem as it is a diplomatic or political issue. Much of defense research is how best to subdue (kill) our enemy or to protect our armed forces. Given the nature of the world, this is highly important. Our survival depends upon keeping the "upper hand" against the "bad guys" of the world.

Research in nuclear physics has both defense and societal relevance. The effort to weaponize a nuclear reaction, the Manhattan Project, is among the most important research accomplishments of the 20th Century. Its success brought an abrupt end to one of the greatest conflicts in the history of mankind. Research continues to develop nuclear weapon technology. There are other nations that are in various stages of becoming "nuclear."

Much nuclear research today is directed at addressing the energy challenge. This process holds great promise as one of the keys to achieving energy security for the planet. From these perspectives, nuclear research holds an important position in the future well-being and security of society.

Space exploration and research have captured the excitement and enthusiasm of millions on the planet. We were enthralled when man first escaped the bonds of earth and years later were mesmerized when man first stepped on the surface of the moon. School children marveled when recently the high quality pictures of Mars started reaching earth.

Much of the research carried out as part of the space exploration effort has become quite useful in agriculture. For example, GPS (Global Positioning System) technology pioneered by space research has become extremely useful in guidance systems for farm equipment. Such systems enable more efficient use of inputs, e.g. water, fertilizer and pesticides. Another area where space research has found uses in agriculture is the employment of ultra-spectral spectrometry employed in earth sensing satellites. Such systems have enabled farmers to keep track of weather in the short term and to monitor climate changes in the long term. Space exploration and research have made discoveries and have led to many developments that have had a positive impact on society.

High energy, particle physics and related research is another exciting research endeavor. Research that focuses on the very smallest particles of matter or radiation helps to explain the essence of our universe. Such research may one day fully define all aspects of matter, antimatter, dark matter, black holes and all the phenomena, which may explain the origin of our universe. Such research is important—even if much of society doesn't understand or appreciate the details. Keep in mind the lay public did not understand and was not universally appreciative of Galileo's research.

Recent research discoveries by physicists continue to challenge the average person and impress many learned scientists on the planet. The discovery of the Higgs boson particle by physicists working at the Large Hadron Collider in the summer of 2012 confirms the theory developed by Peter Higgs in 1964. Reflecting on the importance of this discovery, the identification of the Higgs boson particle was named the most important breakthrough of the year by *Science* (Cho, "The Discovery of the Higgs Boson"). Like so many of the "big" challenges in science today, more than 10,000 scientists work at the European Organization for Nuclear Physics which operates the Hadron Collider. Since many nations, including the United States, have contributed scientists, researchers, and financial support it is a global achievement.

There are literally countless areas of research underway in many forms at many institutions and corporations. Any research that is based on sound principles of scientific investigation that holds interest for other scientists and a significant part of our population is important. Research that explains, enlightens, informs, or leads to a better understanding of ourselves, society/world, or our universe has merit. Consequently, one cannot argue with the importance of any legitimate, sound scientific endeavor. If research meets the criteria as outlined here, it has some societal value. It's simply a matter of priority and where it ranks in attention for funding. If research meets these criteria, at the very least there is one person interested in such research—the person doing it.

Then the obvious question is "why is agricultural research the most important research endeavor?" First, we should recognize that our only source of nourishment and much of our energy sometime in the future will be dependent upon the success of agriculture. The success of agriculture is greatly dependent upon information, knowledge, and technology gained through agricultural research.

All research endeavors contribute to the well-being of our society and civilization. However, agricultural research is the key for the future of our civilization as we know it. With this fact in mind, one can easily see where agricultural research should be positioned in relation to importance. In this book we will expound on the validity of this position.

It is becoming increasingly apparent that research in the agricultural sciences is the nexus of health, nutrition, and food. As more is learned about these relationships, there will be opportunities to design foods with specific desirable qualities for the benefit of humankind. Such developments could enable farmers to produce nutritious foods with specific attributes that could have a positive impact on human health.

Lack of commitment for support of research is, indeed, serious; however, the lack of appreciation of the realities facing the planet are equally, or perhaps even more serious. Developing sustainable energy is not an option—it's a must. The world has operated in and based a vibrant civilization on a finite energy source. While oil discoveries may have peaked, the recent technological advances in energy recovery and natural gas discoveries have increased availability of an important energy source. While still a finite resource, the new sources have influenced the profitability and incentives for investments in alternative energy. This change highlights the uncertainty in energy markets, which represents one of the major challenges for energy investments. However, they do not change the dynamics. Expanding populations along with profligate use of our remaining energy resources hasten the day of reckoning.

There is a lack of appreciation by many of the world's people for new and innovative technologies that provide important ways of addressing food and energy challenges. For example, use of biotechnology in many aspects of research help solve some seemingly insurmountable problems, and yet such techniques are not universally accepted.

Such phenomena as climate change are sometimes not taken seriously. Even though much is not known about climate change, there is ample evidence such changes are occurring. Consequently, there must be some plan to address the impacts of such changes.

One of the major concerns is the lack of appreciation, by some people, of the important roles science plays in helping meet the challenges that lie ahead. Also, there is often a failure to recognize and appreciate the assets we have to help meet these challenges. Evidence is the failure to support research in general, and agricultural research, in particular. The commitment to research to produce a positive future is essential even though the avenues that will be successful are not yet fully evident.

Addressing the topic *Feeding the World: Agricultural Research in the Twenty-First Century* presents a genuine challenge. Having difficulty in deciding just where to start I took the easy path and started with man's observation of nature. Human beings, even with somewhat limited cerebral capacity, could learn much by observing his surroundings. Certainly, he could see animals eating plants while other animals were eating animals. Apparently, it was a small step to emulate the action of the animals. Formal agricultural research is a much more recent development.

Early Experimentation and Emergence of the Land Grant System

Some of the earliest experimentation has its roots in ancient Egypt and Mesopotamia (Grant, "When did Modern Science Begin?"). In fact, until around 1500 such fields of science as mathematics, astronomy, geometric optics and medicine were more highly developed than in the West. The Chinese also had made significant accomplishments in science: however, it was not institutionalized. It was left to the Europeans to institutionalize the research process. Universal education or providing an educational experience for the common man was an innovation of the United States of America. This enlightened contribution made by the US deserves special recognition in the history of civilization. After education, research quickly followed. Because agriculture was a major component of New World America, it is not surprising that agricultural research became an early development.

Agriculture is a dynamic enterprise and, consequently, agricultural research must adapt. Indeed, agriculture has undergone numerous paradigm changes which are fully described in chapters four and five. Success of past agricultural research is evident based on the fact that the world's population has already passed seven billion and overall quality of life continues to improve. The central thesis of this book is to recognize this fact and to plan for a future where population continues to grow and personal expectations continue to increase. India and China have added 88 and 19 million people, respectively, in the period 2009–2012 (Population Reference Bureau, 2009 and 2012). By 2050 population of India is projected to grow to about 1.7 billion surpassing population of China by over 350 million.

This, of course, brings up the question of financial support. Cost of research and return on investment are relevant considerations. A range of other topics are included that bear on the research enterprises including opinions, ideas, and assessment by scientists, stakeholders, politicians, and the consumer.

In a competitive environment, there is the age old issue of public vs. private support of any endeavor. Agricultural research should be considered as a public and private effort because both approaches are complimentary, and as the world becomes more of a community, agricultural research will benefit greatly by international cooperation.

Finally, what are the challenges that lie in the future if our civilization is to prosper? Indeed, what does the future hold for agricultural research? A crucial question is, "is the current agricultural research system the best system to assure success in the future?" The short answer is "no." Even though our system has been phenomenally successful and has met the needs of the planet's population in the past, there is a critical need for changes to ensure success in the future. Later chapters in this book will provide some ideas for strengthening this important enterprise. Among those changes that will be discussed are the creation of a stand-alone agricultural research agency, encouraging more visionary high risk research, recruiting the best and brightest individuals for agricultural research scientists, and improving funding and better funding mechanisms for agricultural research.

A recent positive development is the report by the President's Council of Advisors on Science and Technology (PCAST) released in December 2012. This document entitled, "Report to the President on Agricultural Preparedness and the Agricultural Research Enterprise" illustrates the importance of agriculture in meeting the challenges in the twenty-first Century along with the requirement of a renewed and robust commitment to research, innovation and technology development.[6] To this author's memory, this is the strongest ever public recognition and statement by this important group to relate the importance of agriculture and agricultural research for the well-being of our country and our civilization.

In the foregoing paragraphs are some of the issues this book will address.

1
The Challenge for Agriculture and the Emergence of Agricultural Research

The first essential component of social justice is adequate food for all mankind.
—NORMAN BORLAUG

Food Challenge

Agriculture must provide nutritious food for people and address their expectations of comfort. At the dawn of formal agriculture, the world population was estimated to be between one and five million. While the exact figure is debatable, it is obvious that life was hard for early human beings. In the short span of 10,000–12,000 years, as agriculture evolved and improved, the world's population grew to about 200 million by the year 1 AD. In 1804 it reached one billion; in 1927, two billion; in 1960, three billion; in 1974, four billion; in 1987, five billion; and in 1999, six billion (*Human Population Dynamics: World Population Growth through History*). In 2015 the world's population reached 7.25 billion (ibid.; US Census Bureau World Population Clock). Projections are that the number of people on Earth will exceed nine billion well before 2050. This means there will be two billion more mouths to feed in fewer than 40 years. That is like adding the population of two more Indias or adding the population of China to that of the United States, Japan, Indonesia, and Brazil. In addition to increasing population, most countries are projected to see the population that is at least 65 years of age surpass the group that is younger than 15 by midcentury (Pew Research Center).

World population has increased dramatically since 1950. However, it is encouraging to note that the 10-year growth rate is still decreasing rapidly. In 1950 that rate was 18.9%; 22% in 1960; 20.2%

in 1970; 18.5% in 1980; 15.2% in 1990; 12.6% in 2000; and 10.7% in 2010. Projections indicate that growth rates will be 8.7% in 2020; 7.3% in 2030; and 5.6% in 2040 (Total Population of the World by Decades, 1950–2050).

Meeting both the food and energy challenges for such an expanding population requires judicious use of land resources (*Solutions from the Land*). Wide discrepancies exist in the population densities of the nations of the world. Three of the most populous countries—India, China, and Indonesia—have population densities of 933, 361, and 316 people per square mile, respectively. However, two of the most populous nations, Brazil and the United States, have 58 and 83 people per square mile, respectively (*Countries of the World*)

Arable land is defined as land capable of producing crops. It is further defined by the Food and Agriculture Organization (FAO) of the United Nations as land planted in temporary crops, pastures, gardens, or land for these uses but currently lying fallow. Of the earth's 57 million square miles of land, only about 12 million square miles are arable (*Solutions from the Land*). This translates into 2.56 acres per person in 1959 and 1.15 acres per person in 2006. Assuming a population of 9 billion in 2050, there will be 0.85 acres per person. Although these figures are disconcerting, they do not take into account that arable land is being lost at an alarming rate for other uses such as cities, towns, and roads. These losses, coupled with potential losses due to rising sea levels in response to global climatic change, portray a challenging future. Availability of arable land, along with increasing population and growing expectations and demands, means that the agricultural system will be seriously challenged to meet the future needs of the people on this planet.

As already mentioned, in addition to population growth, people's expectations are also escalating. While this is obviously true for developing countries, even people in developed countries such as the United States want more. As countries prosper, one of the first goals is an abundance of relatively inexpensive food. People also want a better diet, one that includes more fruits, vegetables, meat, milk, and other animal products. Aside from the growing expectations of the existing population, providing the amount of food needed for another 2 billion plus individuals will be a difficult task.

The provision of food for humans, feed for livestock and companion animals, fiber for clothes, as well as homes and flowers for our landscape and environment, is the domain of agriculture and forestry. What is less obvious to some is that agriculture also has the potential to satisfy a portion of the planet's future energy needs. Providing energy is only a start. Plants are able to produce a myriad of products and compounds, including biochemical precursors that could replace or augment hydrocarbon precursors for polymers, pharmaceuticals, cosmetics, and many other chemicals necessary for our society.

Energy Challenge

Our food future is based almost exclusively on agriculture with modest help from aquaculture and mariculture. Agriculture will also play a key role in a future based on sustainable energy. While hydraulic fracturing (fracking) has dramatically increased supplies of petroleum and natural gas in the short run, it has not changed the fact these fossil sources of energy are finite. As we move to a more sustainable energy future, we will become more dependent on the energy from the sun.

The United States is, by far, the world's leading consumer of oil. World consumption of oil was around 93 million barrels/day in 2010 (*World Factbook*), while the US consumption was about 19 million barrels/day. We in the United States use almost as much energy to keep comfortably cool in summer and warm in winter as the total annual energy consumed by all of the countries in Africa. Energy use (kg of oil equivalent per capita per year) is more than 10,000 for several countries, including Iceland, Kuwait, Qatar, and Trinidad and Tobago. Consumption in the United States is slightly more than 7,000, whereas in China it falls below 2,000 and in India below 600.

Other sobering statistics include about 250 million cars, trucks, and buses in the United States. According to the World Bank, the United States had 797 vehicles per 1,000 people in 2009. Canada had 607 motor vehicles per 1,000 population (World Bank Data). China had 233 million vehicles in 2012 (*China's Motor Vehicles*), while India had about 100 million vehicles (India General Data). However, predictions are that India will have 450 million vehicles by 2020. Nobody wants to walk; everybody wants their own personal transportation. The truth is, the United States has less than 5% of the world's population, yet we consume almost 25% of the world's energy. Most people agree this is not sustainable.

Linkage of Agriculture and Energy

How did agriculture come to play such a central role in our survival on this planet? In the preface, Boysie Day is quoted as saying that "Agriculture is not just the most essential industry; it is the only essential industry." This is because it is totally responsible for supplying the food that sustains us. It has been well documented that success in agriculture is greatly dependent on information, knowledge, and technology—the products of agricultural research, which thus is vital. In addition to providing for our food needs, agriculture and forestry will also furnish a portion of the planet's future energy needs. Food and energy are closely linked and are based on capturing energy from the sun. Why the sun? Consider the following three reasons:

1. The sun will be with us for a long time.

Our sun is the center of our solar system and is estimated to be about 4.5 billion years old. Although precisely how our sun (a star) came into being is unknown, it is thought to have formed along with the rest of our solar system from the collapse of a giant nebular cloud of matter. Once the pressure and density of hydrogen became great enough, thermonuclear fusion began.

Astronomers further indicate that the sun will begin to change in the next 2–3 billion years, eventually turning into a yellow dwarf, then into a red giant by the time it is 10 billion years old. This suggests that, even though the sun, our energy source, is not permanent, it is consistent and is unquestionably our best hope for this planet's energy future.

2. The sun is a very powerful source of energy.

Sufficient energy from the sun reaches Earth in one hour to supply the needs of the planet for a year. The total power needs of the humans on Earth are about 16 terawatts a year. (A terawatt is a trillion watts.) In the year 2020 it is expected to grow to 20 terawatts. Earth receives about 120,000 terawatts of sunshine a year. Of course, Earth uses this energy to add warmth and produce biomass, and a portion of it is reradiated back into space. Yet, given the magnitude of the energy hitting our planet, it is clear that using solar energy is a good way to capture power for human needs, and we can collect and store that energy in a number of ways.

3. The sun is a ubiquitous source of energy.

The sun's energy is available to all nations. It can be captured to grow plants or be converted by photovoltaic cells for other energy uses in regions where plants cannot survive, such as deserts. Because of its availability, solar energy would be unlikely to stir up conflict, in stark contrast to the struggles that have arisen over other, more limited energy sources. Another important reason for using the sun as an energy source is that no one owns it. There are no flags of conquest on this star.

To claim its share of the sun's energy, each nation must simply use its land/water area for collection. Unlike fossil sources of energy, as well as wind, geothermal, and hydro energy, with which only certain nations are blessed, every country in the world can count on the sun's energy. The only requirement is that they develop the means to capture the sun's energy in a readily usable form.

Capturing the sun's energy dates to the beginning of our universe. Green plants, through the photosynthetic process, absorb the sun's energy and turn it into something useful. Millions of years ago this process yielded petroleum, coal, and other forms of fossil energy we find so useful today. Now, agriculture and forestry are aiding us in the capture of energy. Farmers are experts in the production of crops, which make use of the sun's energy for the production of biomass, which in turn can be used for food or converted into a more useful form of biofuel or bioproduct.

The Initiation of Research

Even though agriculture began some 10,000–12,000 years ago, definitive, formal agricultural research is a much more recent phenomenon. Agricultural research was built on many earlier developments in the evolution of the concept of research. The observations of early human beings and later scholars such as Aristotle, Abū Alī al-Ḥasan ibn al-Haytham, aka Alhazen, Copernicus, Galileo, Newton, Kepler, and many others gave rise to research.

The scientific method, which provides the basis for research, begins with a hypothesis that can be tested or evaluated. Once a hypothesis is formed, the challenge is for the researcher to design experiments that will either confirm or disprove the proposition. This is, of course, the heart of the scientific method.

Modern science requires the institutionalization of findings. In the earliest days of observation and experimentation as practiced by Muslim schools, scholars had no systematic means of sharing their observations or even the dates of their scientific inquiries. About the only means of communication available to them was personal contact at meetings arranged by scientific societies.

This changed with the formation in 1662 of the Royal Society of London for Improving Natural Knowledge. The society was granted a charter to publish the *Philosophical Transactions of the Royal Society* in 1665, thus becoming the world's first scientific journal. Even though quite different from the scientific journals of today, it served as a means of informing society members and others of the latest scientific discoveries.

A central lesson of science is that to understand complex issues (or even simple ones), we must try to free our minds of dogma and to guarantee the freedom to publish, to contradict, and to experiment. Arguments from authority are unacceptable.

—CARL SAGAN

Publication of results of scientific investigations is important and in fact constitutes the final step in the research process. Failure to publish such results is a failure to complete the research. Publication in appropriate scientific journals has been an expectation for 350 years. Today, the means of communicating the results of research may involve oral disclosure, print media, or electronic methods. Of course, communication helps build understanding of and appreciation for research.

Peer review of science soon followed and today is considered the only legitimate basis for the evaluation of science. Of course, that did not prevent churches, religious groups, governments, and others from weighing in on scientific matters. But today, peer review is considered the only valid arbiter of scientific quality. While religious institutions often have opinions about various aspects of science, they make infrequent efforts to change or influence scientific findings. Unfortunately, even nowadays politicians who are unqualified to evaluate the science sometimes make forays into the field.

Probably the most famous example of governmental or political interference in science was the Russian Trofim Lysenko (Graham, *Science in Russia and the Soviet Union*). Lysenko was an agronomist who rose from exceedingly humble beginnings to become director of the Institute of Genetics within the USSR's Academy of Sciences. His ill-founded concept of genetics, based on anti-Mendelian principles, along with his theories of environmentally acquired inheritance, were eventually completely discredited by the scientific community of the USSR. His rise to power coincided with his grandiose proposal for agricultural productivity, which had no scientific basis, along with his admiration for powerful politicians. This enabled Lysenko to develop a power base of his own for a time.

The rigors of peer review and the validity of the scientific process are critical to ensuring quality in science. However, governments—even democratic societies—still interfere in the scientific process occasionally. Such intrusion is usually driven by environmental considerations or perceived ethics violations. Unfortunately, governments usually address these situations by limiting funding.

Interference by religious entities is almost exclusively along ethical lines. Since these institutions usually provide only a minor source of the funds necessary for scientific studies, their impact on the direction of the research is limited. Probably the greatest interference today is that by individuals or groups who have some particular interest or ax to grind.

Today, celebrities abetted by various forms of news media exert great influence in some areas of science, certainly in agriculture. A number of active groups question safety and the use of many technological advancements, such as herbicides, growth regulators, antibiotics in animal feed, and biotechnology. Other groups point to global climate change. While such groups often mean well, it is imperative that agricultural research provide definitive findings that allay all fears of potential deleterious effects of such technology.

Agricultural scientists are hampered in responding to antitechnology claims for several reasons. Such groups have often utilized the power of the news media by enlisting the assistance of popular movie stars and television personalities who often have loyal and dedicated support groups. However, many such individuals are two or three generations removed from the farm and have little appreciation of what is involved in ensuring our food supply. Finally, it is often difficult to develop definitive answers to questions from the antitechnology groups.

Such groups and individuals who challenge or otherwise interfere in the scientific process are sometimes science challenged but sincere in their concern. The agricultural science community has responded with patience and tolerance for differing views. While its response should be tempered with restraint and forbearance, it must also hold fast to the principles of sound science. The challenges that are facing agriculture in the decades and centuries ahead and the commitments that must be made if it is to be successful must enable agriculture to take advantage of reasonable and sound principles of science and technology.

The challenge usually arises when it is difficult to develop specific experiments to test a single hypothesis. This often occurs when multiple interactions take place, as reflected in many biological systems. In such situations it becomes exceedingly hard to measure the effect of a single factor. The danger posed by pseudogenetics, pseudoecology, and indeed all pseudosciences is of great concern in maintaining agricultural productivity (Borlaug, *Agricultural Science and the Public*).

The communication of scientific findings followed a traditional path throughout the second millennium. It was not until the dawn of the third millennium, the twenty-first century, that the dissemination of these discoveries would undergo yet another major revolution. We are rapidly moving toward the electronic disclosure of scientific information through ejournals and social media such as Facebook, Twitter, and other electronic modes.

Strict adherence to scientific principles will help ensure the success of a research effort. Shoddy science can lead to erroneous results that not only cost time and money but may also substantially delay the finding of a real solution to any given problem.

This brief discussion illustrates several important facts. First, the evolution of the scientific method was made possible by the efforts of many critical thinkers. Second, progress did not occur in a straight line. The evolutionary process has been characterized by many twists and turns. Finally, the foundation was laid for a future where humankind as we know it could not have survived without the rigor of the scientific method.

2
Birth of the Land-Grant College and Its Impact on Agricultural Research

*The 1862 Morrill (Land-Grant) Act changed forever
how higher education is perceived in this country. This
legislation unleashed a hunger for learning by ordinary,
average people that has not been fully satisfied to this day.*

—GALE BUCHANAN, DEAN AND DIRECTOR,
UNIVERSITY OF GEORGIA, 2005

The Land-Grant Concept: A Revolutionary Idea

By leading to the development of the land-grant colleges, which focused on agriculture and the mechanical arts, the 1862 Morrill Act contributed greatly to the success of this country. With the formation of these institutions, civilization—for the first time in history—encouraged all of society's members to seek opportunities to reach their full potential. Over time, that encouragement greatly improved society's intellectual potential. Even more important than this was the philosophy of the land-grant concept.

As these institutions gained credibility, it became apparent that in order to provide meaningful instruction, more information, knowledge, and technology, which could be gained only by enhancing research, were required. Around the mid-nineteenth century there was discussion across the breadth of this country about the need for ways to improve our society. We had long ago become a free nation after severing the bonds with England and fought two wars to ensure that freedom. That left the new nation to its own devices to succeed. It was fortuitous that many visionary leaders emerged who saw clearly how we could do so. They acknowledged that building a successful national economy required three things: (1) building or manufacturing something one needed, could use, or could sell at a profit, (2) growing something one could eat, use, or sell at a profit, and (3) mining or harvesting something from nature or the land. This was a blueprint for building an economy and hence a nation.

There are some who consider the great success of the American land-grant colleges and universities as the culmination of a carefully considered plan. I do not completely agree. On the contrary, early founders took advantage of each situation and made what turned out to be astute decisions. They started with three premises on building an economy and proceeded from there, recognizing that agriculture and the mechanical arts (manufacturing) were clever and realistic first steps. Using this as the basis, everything else fell into place.

Another opportunity was the availability of land. Land—good land for farming or growing timber—was available almost everywhere. Even more was added to the developing country with the Louisiana Purchase in 1803, treaties with Great Britain in 1846 and Mexico in 1848, the Texas Compromise in 1850, the Gadsden Purchase in 1853, and finally the Alaska Purchase in 1867. In the mid-nineteenth century, Congress used its greatest asset—land—to develop and promote education. Congress used land assets in three ways: (1) to improve the land by building highways, railroads, canals, and drainages; (2) to provide grants to groups such as political or religious exiles and veterans of the American Revolution; and (3) to educate all the people. This latter category of land grants has had a profound effect on American education. The first types of land grants were known as section grants. Initially, the sixteenth section of each township was set aside for grade schools. Later, the allocation was increased to two sections and then to four. These section grants provided support for the primary (first- through eighth-grade) schools, which were prevalent throughout the United States early in the twentieth century. These were small, one-room schools that gave every child an opportunity to obtain an eighth-grade education. Because the emphasis was on basic skills (e.g., reading, writing, arithmetic), the result was a literate society. The second type of education grant, known as the township grant, was for the establishment of a seminary and a state university in each state. For a number of states, the seminary often became its first state university. For example, Louisiana State University, the University of Michigan, and the University of Wisconsin all started in this manner.

The Magnificent Charter, a book on the 1862 Morrill Land-Grant Legislation, was authored by J. B. Edmond. Obviously he held this legislation in high regard. To my mind, it is the most enlightened law ever created. It got just about everything right, including the importance of agriculture and manufacturing for building an economy. Also, providing an educational opportunity for the common people as opposed to only the wealthy was a stroke of genius. This latter point might not be thought of as a big deal. Quite the contrary, it was, indeed, a very big deal. In fact, it was revolutionary.[1]

In the old world, if you were a laborer—you, your children, and your children's children would probably die as laborers. Manual workers could hardly dream of an education and a better life. Imagine a sharecropper's son or daughter going to study at a university and then becoming a professional. It would have been unheard of at the time. The Land-Grant College Act probably did more for the average person than any other bill in history. Providing an educational opportunity for the average person meant that those with ability and a willingness to work could enhance their status in society. A sometimes overlooked provision of the act was that it provided for training for military officers. Therefore, it became possible for the son of a "common man" to become not only an officer but even a general.

The land-grant concept was beginning to catch fire. Several individuals deserve recognition for promoting this revolutionary concept. Simon DeWitt of New York, Alden Partridge of Vermont, and Thomas G. Clemson of Pennsylvania were all influential in advancing the land-grant concept and movement. Jonathan Turner, who was born in Massachusetts but spent much of his life in Illinois, laid out a blueprint for the land-grant concept. Turner is one of the unsung heroes of the land-grant movement. Through his effort, the Illinois legislature petitioned Congress in 1853 to donate land that could be sold and to place the proceeds in an endowment for university support.

It would make a good story to say that endorsement of higher education in agriculture garnered universal support, but this would overlook some critical opposition. Certain people felt that agriculture did not need such an educational system. They believed an apprentice system, such as that practiced by many trades, was all that was needed. Fortunately, the educational approach won out. In 1857 Congressman Justin Smith Morrill introduced his first Land-Grant Bill—HR 2—in the US Congress.

This bill was to provide for "the endowment, support, and maintenance of at least one college in each state where the leading object shall be, without excluding other scientific and classical studies, and including military tactics, to teach such branches of learning as are related to agriculture and the mechanical arts, in such manner as the legislatures of the states may respectively prescribe, in order to promote the liberal and practical education of the industrial classes in the several pursuits and professions in life" (First Morrill Act). Congress passed the bill. Unfortunately, as author of this book, I take no pride in noting that the proposed legislation was vetoed by then president James Buchanan.

While President Buchanan was not one of our stellar presidents, his action provides ample evidence that not all politicians or even presidents are smart or visionary. Here is an excerpt from his veto message:

> I shall now proceed to state my objections to this bill. I deem it to be both inexpedient and unconstitutional.
>
> This bill has been passed at a period when we can with great difficulty raise sufficient revenue to sustain the expenses of the Government. Should it become a law the Treasury will be deprived of the whole, or nearly the whole, of our income from the sale of public lands, which for the next fiscal year has been estimated at $5,000,000.

For President Buchanan these were extraordinarily trying times. South Carolina was on the verge of seceding from the Union, and all indications were that, if that happened, civil war would follow. Of course, the legislation was written such that states participating in an act of rebellion could not benefit from the provisions of the legislation. Thus, President Buchanan put an end to the land-grant issues for the time being. But Congressman Morrill had other ideas. The land-grant bill was reintroduced in 1861 as HR 138.

By this time, the land-grant concept had attained even greater credibility and was regarded as having considerable merit. Consequently, after Congress passed the act, President Lincoln signed it into law on July 2, 1862. This was, of course, a critical time for our country, which was engaged in an extremely bloody civil war. Apparently, there was little doubt the president would sign the bill. As the story goes, both candidates for president, Lincoln and Douglas, were visited prior to the election to solicit their support for the land-grant bill should Congress pass it again. So, the land-grant legislation was guaranteed executive endorsement regardless of who won the 1861 election. All the supporters had to do was to ensure that Congress approved it.

Abraham Lincoln was arguably our greatest president. Certainly, most would agree he is among the three greatest presidents—if not the greatest. He was truly a man of the soil, who was at home doing any of the chores common on the typical pioneer farm. He was an expert with the ax and a master rail-splitter. Of course, as president, he worked diligently to preserve the Union and free the slaves, forever changing this country. Lincoln is also responsible for other major decisions that shaped agriculture in the United States. The Homestead Act did much to open up our country and to increase the productivity of the land. He signed legislation that created the land-grant colleges, changing higher education forever. Moreover, he also created the Department of Agriculture (USDA), which he called the "People's Department." Today, the USDA provides leadership for farm programs, conservation, forestry, marketing of agricultural products, food support for the needy, and agricultural and forestry research. Abraham Lincoln was indeed an agricultural president!

Land-Grant Institutions Become a Reality

The particulars of the land-grant legislation were quite straightforward. First, public land was to be apportioned to each state in the amount of 30,000 acres for each senator and representative in Congress. If a state was short on federal land, it would receive scrip equal to federal land in another state or region. Specific requirements governed investment of the money resulting in the sale of the land or scrip. Also, no state, while in a condition of rebellion or insurrection against the US government, would be entitled to the benefit of this act. This meant that none of the Confederate states could apply until hostilities ended and they were accepted back into the Union.

While the Federal Land-Grant Act provides the basis for the land-grant university, the Morrill legislation specifically addressed only a part of its mission: teaching. Rounding out the tripartite mission of the land-grant university, which has come to be thought of as the land-grant concept, are outreach and research. Sometimes this concept is referred to as a three-legged stool. The implication is that all three legs are critical, and if any one of the legs is missing, the effectiveness of the stool is minimized.[2]

The tripartite nature of agricultural programs in today's land-grant universities has been copied by other components of the university. In the typical land-grant university many colleges and departments have strong teaching and research programs. Some departments also have effective outreach or extension programs. While decidedly different from cooperative extension in agriculture, these programs add an outreach dimension to the particular department.

Cooperative extension, which serves as the outreach arm of agriculture, has the distinct advantage of federal recognition and support. Because of this, there is a national system that adds great strength to the effort. Most outreach efforts in other parts of the university do not have this advantage. The land-grant concept is also often emulated in other parts of the world; however, few countries have been as successful as the United States in molding agricultural research, extension, and teaching into a totally cohesive unit.

The land-grant university's impact on research is often overlooked. The Morrill legislation required that agriculture and mechanical arts be taught without excluding other scientific and classical studies. Unfortunately, at the time a clearly defined scientific basis for these fields was lacking.

Nevertheless, some of this nation's most cherished visionary leaders were also farsighted agricultural innovators. Both Thomas Jefferson and George Washington routinely experimented with new varieties of crops and with cultural practices on their farms. Teaching agriculture when a void in the science exists is simply not possible.

Consequently, it was absolutely imperative that something be done to create a science base in order to teach scientific agriculture. As educational programs continued to develop, the need for more information to teach became clear. Thus, the legislation that created the land-grant institutions provided a great stimulus for agricultural research.

While the Confederate states were temporarily prevented from participating in the Morrill legislation, African Americans were completely disenfranchised. To rectify this intolerable situation, legislation was passed in 1890 to create an additional set of land-grant institutions. This legislation is often referred to as the 1890 Act or the Second Morrill Act. The bill was intended to provide educational opportunities for African Americans. Although the initial assistance was modest, over the years it has increased. Federal support for a research mission in the 1890 institutions was added much later.[3] However, both state and federal support for these 1890 land-grant universities (LGUs) still lags behind that for the 1862 LGUs.

Tuskegee University, one of the 1890 institutions, is a private institution located in eastern Alabama. Founded in 1881 as the Tuskegee Normal School for Colored Teachers, it is today a thriving land-grant university. Early leadership by President Booker T. Washington was instrumental in the success of the institution. One of his most important hires was the notable George Washington Carver. Today the agricultural experimental station at Tuskegee is named in his honor. The Tuskegee Normal School has undergone several name changes, including Tuskegee Normal and Industrial Institute and Tuskegee Institute, and today is known as Tuskegee University. The most recent addition to the land-grant story is the 1994 legislation, which provided support for selected Native American (primarily two- and four-year) institutions of higher learning.[4]

It is remarkable how each land-grant college or university has evolved. Some LGUs are now listed among the elite universities in the land. Among these are Cornell, the University of Wisconsin, the University of California, Texas A&M University, and the University of Florida. Other outstanding LGUs fully and proudly embrace agricultural teaching, research, and outreach. Chartered as a state university in 1785 and initially offering a liberal arts education for the privileged classes, the University of Georgia has today become one of the nation's outstanding land-grant universities and boasts a strong College of Agricultural and Environmental Sciences (Wade, "Intent and Fulfillment").

The truth is that each of the nation's LGUs is a special institution because each excels in some particular area and in its own unique way. Even though the various land-grant institutions show great disparity in size, quality is not necessarily associated with size; however, smaller institutions are simply unable to have the same number of quality programs as do the larger institutions. The bottom line is that each of these institutions has graduated many of the people who are responsible for making this country what it is today.

3
Institutionalization of Agricultural Research

*I know of no pursuit in which more real and important
services can be rendered to any country than by improving
its agriculture, its breed of useful animals, and other
branches of husbandman's cares.*

—GEORGE WASHINGTON

Early Developments in Science

In the seventeenth, eighteenth, and nineteenth centuries, research
led to a number of discoveries that fostered the beginning of scientific agriculture. Some of these include Robert Hooke's discovery of
the cell in 1635; the law of conservation of mass, which provided a
basis for chemistry, and the work of Antoine Lavoisier, which led to
the inception of modern chemistry in 1789; Charles Darwin's 1859
theory of natural selection; Louis Pasteur's 1861 germ theory; and
Gregor Mendel's 1865 work on the laws of inheritance, which provided a basis for genetics. Another fundamental contribution was
Niels Bohr's development of a better understanding of the nature of
the atom, atomic structure, and quantum mechanics.

The identification of "nuclein" (nucleic acid) by Friedrich Miescher
in 1869, followed by the research of Oswald Avery, who in 1943
proved that one of the nucleic acids, DNA, is the genetic material of
the chromosome, and finally the determination of the helical structure of DNA by J. D. Watson and F. H. C. Crick in 1953 were major
discoveries. These developments paved the way for investigations in
other sciences, especially agriculture.

While the documentation of observations and some form of experimentation dates to the first millennium, definitive research was
not institutionalized until the sixteenth and seventeenth centuries.
Accumulation of these developments made possible the concept of

scientific agriculture. Early research primarily focused on the basic sciences and for the most part did not include agriculture. Nevertheless, it is not surprising that agriculture was an early adopter of the scientific approach because the results of scientific inquiry could be used in this area to quickly provide a public benefit. Much of the early experiments and investigations in agriculture were practical and would today be called applied research or demonstrations. Still, even though they were simple, these demonstrations were important and paved the way for more rigorous investigations.

The opportunities for advancing agricultural research made possible by the opening of the New World are often overlooked. In Europe, historical traditions governed agricultural production, but in the New World everything was new. As early colonists arrived in North America, they found already existing there a primitive form of agriculture practiced by Native Americans, which provided for their requirements just as did the agricultural systems of the Europeans. However, the entrance of the English colonists provided a stimulus for change in the New World, including different expectations. Consequently, in addition to the species of plants they found in the new land, such as maize, beans, pumpkins, and tobacco, which were grown by the Native Americans, they also planted other crops they were familiar with (e.g., cereals, such as wheat, barley, rye, and oats; fruit, including apples and peaches; and vegetables, such as turnips and carrots) (True, *History of Agricultural Experimentation*).

One early development that had a far-reaching impact on agriculture was the aid given by governments, corporations, and other groups in support of agricultural research and investigations. As early as 1622, James I of England encouraged the breeding of silkworms (Liebig, *Organic Chemistry*). Later, in 1656, every landowner in Virginia was required to plant ten mulberry trees for each 100 acres possessed. There were numerous other examples during this period of governments' use of various incentives to stimulate agriculture.

Experimental and Demonstration Efforts

In the latter half of the seventeenth century, effort was placed on developing the Carolinas. One noteworthy effort is the Ashley River Settlement. It is memorable primarily because of its provisions for advancing agriculture. This effort included the establishment of a testing garden in 1669–1670. Consequently, this is probably the first agricultural experimental (demonstration) farm established in America for improving agriculture (Carrier, *The Beginnings of Agriculture*).

In 1733, James Oglethorpe was charged with establishing a settlement, which later became Savannah, GA. One of his first actions was to set up an experimental or demonstration garden. In all likelihood the leaders of the settlement profited from the experiences associated with the Ashley River Settlement in South Carolina. The garden was described as follows:

> There is laid out near the Town by order of the Trustees a Garden for making experiments for the Improving of Botany and Agriculture. It contains ten Acres, and lies upon the River; and it is cleared and brought into such Order that there is already a fine nursery of oranges, olives, white mulberries, figs, peaches, and many curious herbs; besides which there is [*sic*] cabbage, peas and other European pulse and plants, which all thrive. (Carrier, *Beginnings of Agriculture*)

Obviously, this was a demonstration garden, and it served nicely to provide useful information for the colonists. The demonstration method usually does not provide definitive, quantifiable results; however, it does supply a great deal of information with only a modest investment. Even under these circumstances this is generally regarded as one of the early agricultural research stations in the New World.

Another note from this report is that olives thrived. That should provide some comfort to those farmers in southeast Georgia and northeast Florida today who, many years later have invested heavily in planting olive trees in the region. There are numerous similar examples of demonstration-type experiments either by individuals or by governments during the early development of this country.

The fate of Oglethorpe's (trustee's) garden was an omen for future support of agricultural research. On a historical sign at the garden site are these words: "The silk and wine industries failed to materialize. The distant sponsors were unable to judge of the immense importance of the experiments/demonstrations that were conducted. In 1755 the site was developed as a residential section."

Farmers were smart then, just as they are now. Constantly searching for ways to improve production, they saw the benefits of experimentation. Although they did not want to use their production capacity to do the research, they were pleased to see others (e.g., governments) invest in research that might yield useful results. They persuaded various societies and governments to do research that directly benefited them and were of indirect advantage to all of society.

We usually think of our first president, George Washington, as a great general, leader, and head of state. He was all that, but, as a leading farmer, he was also highly knowledgeable about agriculture. Today he would be considered a commercial farmer. He read the latest literature on farming and was continually carrying out studies and demonstrations on his farm. Consequently, in his 1790 message to Congress, he suggested it would be well to encourage agriculture as well as commerce and manufacturing.

Washington was greatly impressed by the 1793 establishment of the British Board of Agriculture. Encouraged by Sir John Sinclair, manager of the British Board, and conferring with his colleagues Alexander Hamilton and John Jay, President Washington asked for their "joint opinion" (True, *History of Agricultural Experimentation*). In a presentation in 1796, he extolled agriculture's importance and recommended that the US Congress set up a national board of agriculture. His words were prophetic:

> In proportion as nations advance in population . . . the cultivation of the soil [becomes] more and more an object of public patronage. Institutions . . . grow up supported by the public purse . . . Among the means which have been employed to this end none have been attended with greater success than the establishment of boards composed of proper characters charged with collecting and diffusing information, and enabled by premiums and small pecuniary aids to encourage and assist a spirit of discovery and improvement. This species of establishment contributes doubly to the increase of improvement by stimulating to enterprise and experiment and by drawing to a

common center the results everywhere of individual skill and observation and spreading them thence over the whole nation. (Bishop, *Historical Sketch;* True, *History of Agricultural Experimentation*)

Regrettably, this was not the last time the US Congress did not support a good idea by the chief executive.

This, of course, was only the beginning. In today's parlance, the proposal was "talked to death." However, in 1820 the House of Representatives created a committee on agriculture, and the Senate followed suit in 1825. Unfortunately, these committees did not or were not able to do what President Washington had envisioned.

Linkage of Science and Agriculture

The linkage of science and agriculture grew primarily out of developments in Europe in the early part of the nineteenth century. Among the first areas of science to be linked to agriculture was chemistry. Contributing greatly to this development was Justus von Liebig, a German chemist who offered a number of theories to explain plant growth (Liebig, *Organic Chemistry*). As with many important scientific developments, Liebig's great idea was oversold. Even though his explanation of nutrients and plant growth was an important contribution, the field of soil fertility proved far more complex. There simply was not a single explanation for the relationship between nutrients and plant growth.

The mid-nineteenth century proved to be an exciting time from several perspectives. Developments in science were ushering in a new age. The American Civil War was a major world event. The US political scene was in great turmoil. Slavery, preservation of the Union, and the possible secession of several southern states were extremely contentious topics. Even with this turbulence, the US Congress passed and the president signed several important pieces of far-reaching legislation that had a significant impact on our country, especially on agriculture. One particular bill, the Morrill Act, which has greatly influenced higher education, also had an unexpected consequence: It stimulated agricultural research.

Since scientists in these new institutions usually fulfilled both a teaching and a research function, one can easily surmise that teachers would realize that research would enable them to better instruct students. Of course, any good teacher is always seeking ways and means of expanding the bounds of knowledge. Consequently, a natural synergistic relationship between teaching and research emerged. As outreach evolved, it became an integral part of the process, eventually strengthening the tripartite concept of the total land-grant university. Though communication between the New World and Europe was slow and cumbersome, it was not impossible. Developments in Europe quickly found their way to the United States, where they were modified to fit the American philosophy of how things could or should be done.

Two years after Liebig's 1840 publication, *Organic Chemistry in Its Relation to Agriculture and Physiology*, John Bennett Lawes patented a process for producing superphosphate (*Rothamsted Experimental Station*). This discovery led to the initiation of the Lawes Chemical Company to manufacture this and other plant nutrients in England (ibid.).

Establishment of the First Permanent Experiment Station

In England the Rothamsted Experimental Station was founded in 1843 on the ancestral lands of the Lawes family. Joseph H. Gilbert, a chemist, joined Lawes as a collaborator to develop field experiments to relate crop growth to soil fertility. Thus was born the formal, permanent experiment station as an institution for agricultural research. For many years, Rothamsted Experimental Station was supported by Lawes. However, since the early twentieth century, the station has received assistance from an annual appropriation of public funds. Having visited Rothamsted, I can state that it is a special place for those who appreciate agriculture in general and agricultural research in particular.

Other research initiatives throughout Europe and Great Britain followed the establishment of the Rothamsted Experimental Station. A number of German states, led by the state of Saxony, established agricultural research stations. For the most part, these were highly decentralized and were detached from academic centers.

Soon there were similar developments in the United States and several other European countries. As in Germany, most of the European stations were almost totally decentralized. There were about as many rationales for creating these stations as there were stations. Their purpose was to allow interested individuals to solve specific problems in specific locations.

In the United States, one of the compelling reasons for setting up an agricultural experiment station was soil fertility and the quality of fertilizer being sold by merchants. The discovery of the role of nutrients in plant growth ushered in the commercial fertilizer business. Obviously, this was a marketing opportunity tailor-made for unprincipled salesmen, while at the same time providing great opportunities for reputable merchants. It was also a natural for experiment stations, often in cooperation with the land-grant university teaching programs, to collaborate in establishing the true value of commercial fertilizers. In fact, the initial task of the nation's first agricultural experiment station—the Connecticut station—was to test fertilizer for label compliance. It is an irony of life that unscrupulous merchants were one of the driving forces in the development of agricultural research in this country. This fits the old axiom that "there is often some good even in undesirable acts."

Toward the end of the nineteenth century, the idea for agricultural experiment stations in each state became increasingly popular. It is instructive to note the reasons for such interest. First, agriculture is fundamental to the existence of humankind by providing food, clothing, and shelter. Even at that time, a majority of the population lived on the land. The United States had an abundance of land that was suitable for agricultural pursuits. Almost everyone could recognize that the key to enhancing agriculture was information, knowledge, and technology.

Research became the primary means of developing this useful, new means of making agriculture more productive. Though this clearly improved the economic condition of the farmer, the real beneficiary was the public at large. For these reasons, agricultural experiment stations were ultimately set up in a number of states and were perceived to be a wise investment of public resources. In 1875 one was established in Connecticut, and about six months later that year another was created in California. These were followed by stations in North Carolina in 1877; Massachusetts in 1878; New York (Cornell) in 1879; New Jersey in 1880; New York, Ohio, and Tennessee in 1882; Louisiana in 1884; and Alabama and Kentucky in 1885 (Edmond, *Magnificent Charter*). There is some discrepancy, however, in the order of their founding, depending on which author one consults (ibid.; True, *History of Agricultural Experimentation;* Woodard and Waller, *New Jersey's Agricultural Experiment Station*). For instance, Woodward and Waller (ibid.) wrote that New Jersey was the third state to establish an agricultural experiment station by special act of the legislature. However, they acknowledge that such facilities had begun in three other states as well, though not by legislative enactment. California was established in 1875, Massachusetts in 1878, and New York (Cornell) in 1879. The Massachusetts station was founded in 1878 but ceased to function three years later. The New York (Cornell) station was begun in 1879 and operated for two years without any special funds. The New Jersey Agricultural Experiment Station dates to 1880.

A definitive publication, "A History of Agricultural Experimentations and Research in the United States" by Alfred True, confirms Woodward and Waller's observation that a resolution of the Connecticut legislature of 1875 provided for an agricultural experiment station. True further states that in that same year the regents of the California University System provided $250 for "an industrial survey and experiment station" (True, *History of Agricultural Experimentation*). Apparently some authors considered appropriate legislation a requirement for the formalization of such a facility. What was important and becoming universal was the idea for agricultural research and the institutionalization of such research in an experiment station. By 1887, the year the Hatch Act was passed, fourteen states had some type of an agricultural experiment station. No two facilities were alike. They were, in general, poorly funded. They differed in mission, organization, and just about every other way imaginable. In some instances, clashes occurred with teaching programs in the agricultural colleges.

Challenges for an Experiment Station

Since the whole idea of scientific agriculture, including teaching and research, was new, debate often arose as to how these issues would, could, or should be addressed. For example, administrators of the teaching programs often looked on the research farm as a source of fresh meat and vegetables for the students or, if not as a source of food, then as a model farm for demonstration rather than a field research laboratory. These arguments started early on and, unfortunately, exist to this day at some institutions. Having been an experiment station director in two states, I found that it was almost impossible to convince university administrators and others that these properties were as much a research asset as was a laboratory.

Aside from developing a strong funding base to support agricultural research programs, protecting land assets has always been one of the most challenging responsibilities for the administrative head of these programs (Jordan et al., *Leadership in Agriculture*). After weighing the importance of the land and the strength of the university president, one usually found that fighting about land was not worth the effort unless one could muster the assistance of others who had more political power.[1]

Local governments were equally challenging in expropriating research land assets. Their arguments could be more easily rebuffed— unless they had the foresight, which they usually had, to get the university president in their corner. Their arguments were for a new school, hospital, economic development, or a multitude of other worthy objectives.

Emerging experiment stations had other difficulties. The agricultural community's expectation of solving immediate problems affecting production versus addressing more fundamental issues that had implications for future productivity had to be carefully weighed. Farmers tended to favor the former approach, whereas directors often had a longer-range view. Another serious challenge was the perennial shortage of funds to accomplish the needed and envisioned research to support a robust and growing agricultural economy.

Institutionalization of the Experiment Station

As more states saw the wisdom of having an agricultural experiment station, the need for a more unified system to manage them became apparent. Up to this point each state addressed the need for agricultural research in its own way with widely differing approaches to developing these facilities, all with inadequate funding. In only a few years, it became apparent to agricultural leaders that action must be taken to address the lack of coordination among the existing institutions. Less than a decade after the Morrill Act was passed a major effort was mounted to address this need for coordination (Kerr, *Legacy*). Representatives from a number of agricultural colleges met in 1871. A proposal was submitted by Willard C. Flagg, trustee and director of experiments and superintendent of farms at Illinois Industrial University, to encourage the development of agricultural experiment stations as a part of the land-grant university in each state. The US Department of Agriculture (USDA) followed up the next year with a meeting that included delegates from thirty-two states and three territories. They again proposed the development of an agricultural experiment station in each state. It seemed everyone agreed on the merit of this idea, but no one had a good plan for putting the idea into play.

After an 11-year hiatus, in 1882 a number of the Morrill Land-Grant College presidents and several professors met, resulting in the formation of the Association of Land-Grant Colleges and Experiment Stations (Edmond, *Magnificent Charter*). Although agricultural education was a primary focus of the meeting, considerable interest in agricultural experimentation was also voiced. This was particularly true for those institutions that had already initiated development of an experiment station. As in so many meetings, progress was made in redefining the need; however, it was still not a done deal. In 1885, a delegation from this association led an effort to urge Congress to establish an experiment station in each state (ibid.).

Seaman Knapp[2] proposed that the national treasury provide financial backing for an experiment station at each land-grant college (Kerr, *Legacy*). Frustrated by the lack of support from both the Iowa state government and the federal government to fund the state's agricultural experiment stations, Knapp proposed a joint (state and federal) funding arrangement. Giving birth to the agricultural experiment station system obviously was not an easy task.

For the next five years, there were several efforts to implement the Knapp proposal or some version of it. History should be kind to the visionary members of Congress who supported revolutionary ideas such as agricultural experiment stations, including Representatives Cyrus C. Carpenter of Iowa and William Cullen of Illinois. Momentum continued to build, indicating that something needed to be done. There had been so much talk and the need was just too great not to take the final step, which led to the creation of the state agricultural experiment station system.

The Hatch Solution

It was left to William Henry Hatch of Missouri, chairman of the House Agricultural Committee, to put forth legislation creating the state agricultural experiment station system. In the Senate, an identical bill was sponsored by Sen. James Z. George of Mississippi. This legislation, known as the Hatch Act, created the state agricultural experiment stations (Kerr, *Legacy*). The bill was signed into law by President Grover Cleveland on March 2, 1887.[3] One of the concerns regarding this legislation pertained to the extension of federal power (i.e., federal versus states' rights). More than a hundred years later, this is still a hot topic.

Even though probably not realized at first, the Hatch Act of 1887 was a brilliant piece of legislation. In essence, the federal government provided financial backing, although each state was required to provide matching funds. In addition, an organizational structure was laid out for the system. However, it was left to the discretion of the respective states to design research programs that reflected local needs.

This latter provision was a superb idea that ensured the relevance of research programs, which would be required to meet many local needs; however, long-term needs could be met only through more basic, visionary research. This issue is still a concern today. Clearly, the strength of agricultural research lies in maintaining a proper balance between solving immediate problems and conducting more visionary research that is designed to solve problems that lie in the future.

The Hatch Act has not led to the creation of identical experiment stations in each state. In fact, about the only common denominator is a commitment to conduct investigations that enhance agricultural enterprises in each state. The history of most of these facilities has been featured in books and other materials that provide detailed information about the initiation and evolution of each state's agricultural experiment station.

The manner in which the Hatch legislation was written is unusual with regard to control. Those who supported a centralized form of government found it difficult to understand why federal money should be given to a state with so little direct control. The act clearly states that it is up to the experiment station director to decide how to use the Hatch funds. The director also has the responsibility of setting research priorities. Though some may see this as a weakness, I see it as one of the great strengths of the US agricultural research system.

Where to Locate the Station

The key question was, should the ultimate control reside at the federal level or with the state? But this was not the only bone of contention. Not all states agreed to locate the experiment station within their Morrill land-grant university. For example, the University of Georgia could not or would not make land available for the experiment station. Consequently, land was obtained near Griffin, just south of Atlanta. Initially, the Georgia experiment station was located in Experiment, GA, but the address later became Griffin, Georgia. Following the decision to set up the experiment station at Experiment, Georgia, the Farmers' Alliance made a major push to move the A&M College from the University of Georgia in Athens to Griffin. The effort was unsuccessful (Wade, "Intent and Fulfillment"). A few years after initiation of the Georgia Station in Griffin, leaders in the southern part of the state pushed for and eventually created an experiment station in Tifton, GA. It was not until a major reorganization of all higher education in Georgia that another facility was constructed on the University of Georgia campus in Athens.

In Connecticut, an experiment station was initiated in 1875. For about the first two years it was located at Wesleyan University in Middleton, then moved to Yale University for about five years; however, in 1882 it was relocated to New Haven. With the passage of the Hatch Act the Connecticut Agriculture College, now the Connecticut College of Agricultural and Natural Resources, located in Storrs, also erected an agricultural experiment station on its campus. Since each station had substantial backing, the Hatch money was equally divided between the two stations, which remain separate institutions today.

Ohio presents another interesting story. Like a number of other states, Ohio had created an experiment station in 1882, located on the Columbus campus of the Ohio State University. However, the facility and the university had separate boards of directors. With passage of the Hatch Act and a common interest in a Hatch-supported experiment station, the two groups reached a short-term compromise. However, in 1892, much of the agricultural research for the Ohio station was relocated to a newly acquired farm near Wooster. Today, the Ohio Agricultural Research and Development Center (OARDC) in Wooster serves as a major campus and focuses on agricultural research in Ohio as part of the College of Food, Agricultural, and Environmental Sciences of Ohio State University, with supporting faculty and programs in both Columbus and Wooster, as well as within a network of ten secondary research locations throughout the state.

When the Hatch Act was passed, California was the first and only state in the western part of the nation that had an experiment station. This facility was headquartered on the Berkeley campus. While there was a continuing effort to enhance research programs at Berkeley, the uniqueness of agriculture in California led to an effort to address its specific problems at a number of research sites. Thus, California now has major agricultural research campuses at Davis, Riverside, and Berkeley. Today, many of the institutions of the University of California system carry out some form of agricultural research.

In several states, the original location of the experiment station just did not work out for one reason or another. For example, those originally located at the University of North Carolina at Chapel Hill and the University of Mississippi were later relocated to North Carolina State University in Raleigh and Mississippi State University in Starkville, respectively.

In several states, the initiation and evolution of the experiment station is an interesting story. Kemp Battle, president of the University of North Carolina (UNC) from 1876 to 1891, possessed great interest in agriculture and was active in many agricultural activities and societies (Carpenter and Colvard, *Knowledge Is Power*). After a trip to Connecticut to visit the first US Agricultural Experiment Station and its director, Wilbur O. Atwater, Kemp began advocating for an experiment station at UNC. Even though UNC had control of the Morrill money and the experiment station, the envisioned success was not achieved. Pressure from farmers and various leaders for the development of an agricultural college led on March 7, 1887, to the establishment of the North Carolina College of Agricultural and Mechanical Arts, now North Carolina State University (NCSU). Agricultural programs for North Carolina are now located at this institution. In a close association that exists to this day, NCSU was developed along with the North Carolina Department of Agriculture. An interesting sidelight to how things evolved in North Carolina is that money generated by the imposition of a dog tax[4] was earmarked to support agricultural programs.

Mississippi received its land grant after the state's reentry into the Union in 1870. Income from the sale of the state's land grant was first used to create an agricultural and mechanical program for whites at the University of Mississippi in Oxford and for African Americans at Alcorn University in Lorman. University of Mississippi faculty strove valiantly to get a program going, but students would not support it. After only five enrollments in "agriculture and mechanical arts" in 1873 and three in 1874, the whole program was abandoned in 1876. Later, the experiment station was established at Mississippi A&M (now Mississippi State University) by the state legislature immediately following passage of the Hatch Act. It began operating in 1888. The program at Lorman was successful and continued as Alcorn A&M College. In 1890 this college was designated an 1890 land-grant institution.

Other models include the two agricultural experiment stations in New York: the Cornell University Agricultural Experiment Station in Ithaca and the New York State Agricultural Experiment Station

(NYSAES) in Geneva. The latter was established by the state legislature in 1882 and became a part of Cornell University in 1923. Each agricultural experiment station has a director who serves under the dean of the College of Agriculture and Life Sciences and receives separate Federal Formula Fund allocations.

Emergence of the Agricultural Research Service

Two of the founders of the United States, George Washington and Thomas Jefferson, had a significant interest in agriculture and were leaders in observations and experiments relating to it (True, *History of Agricultural Experimentation*). Their interest and early investigations preceded by several decades the beginning of the formalization of agricultural research.

The enactment of the first US patent law in 1790 and, later, the appointment of Henry L. Ellsworth as superintendent of patents in 1835 provided a means of collecting and disseminating seeds and plants for the purpose of improving agriculture. Though these efforts did not constitute definitive research, they were important forerunners of the research process.

The passage of legislation creating the USDA and the appointment of the first commissioner, Isaac Newton, who had served as chief of the agricultural section of the US Patent Office, sparked interest in and facilitated the promotion of experimental and scientific work. It is of great interest to me that Commissioner Newton was a visionary who saw collecting and distributing seeds and plants, answering questions from farmers and others, along with chemical analysis of soils, grains, fruits, vegetables, and manures, creating professorships, and establishing a library as high priorities. These are, of course, the hallmarks of agricultural research and education (True, *History of Agricultural Experimentation*).

The emergence of state agricultural experiment stations and the rapid development of experimental investigations by the USDA occurred during the latter part of the nineteenth century. At the USDA, such work was carried out by a number of bureaus. The Agriculture Research Administration (ARA) was established in 1942 to coordinate the research functions of the various bureaus of the USDA. Another major reorganization occurred in 1953 with the abolishment of the ARAs and the many bureaus. With this change, the newly created Agricultural Research Service (ARS) assumed the role of the in-house agricultural research agency for the USDA.

Today the mission of the ARS is to develop and publicize solutions to agricultural problems of high national priority; disseminate information to ensure high-quality, safe food and other agricultural products; assess the nutritional needs of Americans; sustain a competitive agricultural economy; enhance the natural resource base and the environment; and provide economic opportunities for rural citizens, communities, and society as a whole. To accomplish this mission, more than 2,000 scientists and 6,000 supporting staff are involved. More than 800 research projects in 17 national programs are endeavoring to meet this challenge. One of the great strengths of the US agricultural research system is the strong in-house research capability of the ARS coupled with the highly distributed research capability of the state agricultural experiment stations. Though differences abound in the state agricultural experiment stations and the ARS, they are very complimentary in addressing the myriad of issues and problems in US agriculture. These two systems, in cooperation with the private sector, form an agricultural research capability that, if properly nurtured and funded, can meet the future needs of the people in this country and contribute to meeting those of people around the world as well.

Multiple Research Sites

Establishment of a state agricultural experiment station in each state was only a beginning. Variations in soils, climate, geography, and crop and livestock enterprises all influenced agricultural production systems. Consequently, each state had a pent-up demand for multiple experimental sites. The more diverse a state's soil and environmental conditions and agricultural portfolios, the greater its need for several research locations. The "main station" generally refers to the research site located at the headquarters of the state agricultural experiment station. Usually this location provides research opportunities for most areas of study at the experiment station. The majority of these facilities have other research sites away from the main station, collectively thought of as branch research sites. These are referred to by various names:

- Agrilife research and extension centers
- branch experiment stations
- branch stations
- fields
- research and education centers
- research and extension centers
- research farms
- research fields
- reserves
- substation farms
- substations

It is interesting to note that the University of California at Berkeley was the first institution to recognize the need for branch research sites (Huffman and Evenson, *Sciences for Agriculture*). Eugene Hilgard, who became director of the California Agricultural Experiment Station in 1877, was an agricultural chemist who recognized California's great diversity of soils, climate, and geography. Using his scientific knowledge and abilities along with his leadership skills, he brought about the creation of branch research centers. Though not all of the centers he established continued to operate, the concept was established. Today, most experiment stations provided for by the Hatch Act of 1887 have additional branch research sites. One such site designated as the university farm for the agricultural branch of the University of California at Berkeley is today the distinguished University of California at Davis.

Personnel associated with these secondary sites initiated the Research Center Administrators Society (RCAS) in 1964. The mission statement of the RCAS simply states that the association aspires "to advance the acquisition and dissemination of scientific knowledge concerning the nature, use, improvement, and interrelationships of research center administration, scientific research, and new technology. To this end, the society shall:

- promote effective research,
- disseminate scientific research,

- facilitate technology transfer,
- foster high standards of education,
- strive to maintain high standards of ethics,
- promote advancements in this profession, and
- cooperate with other organizations having similar objectives"

Research center administrators have a unique position in the agricultural research systems. They live among many of the practical problems that research is working to solve. Consequently, they are a great asset to agricultural research programs.

A Network of Secondary Research Sites

Most states have some branch research sites to support their agricultural research effort. However, these exhibit wide variation in organization and operation. The state agricultural experiment station system has more than 500 such sites. Moreover, the USDA's Agricultural Research Service has almost 100 research sites throughout the United States and in France, Argentina, Australia, and China. The US Forest Service has 80 experimental forest sites and 67 laboratories organized around five regional research stations.

In addition to these aforementioned permanent sites, almost all experiment stations have several temporary research sites located in farmers' fields, forests, ranches, pastures, highway and power line rights-of-way, and so on. From these positions they are able to take advantage of unique soils, fertility levels, geography, pest populations, or environmental conditions, allowing scientists to conduct research where unique situations exist or where invasive species threaten the native fauna and flora. This is an important approach since it would not be wise to establish a population of an invasive species or a crop pest at a main station or a branch research site.

To many, especially those unfamiliar with research, this plethora of branch research sites seems like massive duplication. However, agricultural success is often determined by soil, geographics, and climatic conditions; thus sound, science-based reasons for secondary research locations are numerous. During the past few years many such facilities have been discontinued—not for a lack of need but rather to cut expenses. Most state agricultural experiment stations and the USDA-ARS are in a state of constant review to ascertain whether a particular location is still relevant. Usually, ample justifications exist for continuing each site. Nonetheless, as budgets become tighter, the need to prioritize the sites grows. Closer coordination among state agricultural experiment stations, and USDA-ARS is one means of reducing branch research sites without losing the capacity to conduct site-specific research.

All States Establish Experiment Stations

By 1887, 14 states had established some type of formal agricultural experiment station. By the end of 1888, with support from the Hatch Act of 1887, each state had established some form of research facility. Although these were supported, at least in part, by the state treasury, either from general appropriations or fees, the available funds were often meager, forcing stations to operate on a shoestring. And as one would expect, such diversity among the stations meant they operated and/or reported the results of their research with little uniformity. A mechanism had not yet been developed to ensure that states could compare their approaches (even the particularly desirable ones) or share their results with each other.

Passage of the Hatch Act in 1887 did two important things. First, it provided $15,000 per annum to support each state agricultural experiment station.[5] Moreover, it brought about some measure of uniformity in agricultural research by supplying an operational framework that facilitated routine meetings of experiment station directors and established a unified system for collecting and sharing research results and identifying research priorities. Though these may appear to be minor points, they contribute greatly to a system. Although this framework has taken many forms in the past and is still evolving, today a degree of commonality exists in agricultural research that enables even more effective accountability and sharing of research results.

Within a year, each state had sufficiently established an agricultural experiment station to qualify for its share of the Hatch funds. This was only the first step. How each state used the funds was more problematic. In some states the monies were divided among existing experiment stations that had been initiated separately from the state's land-grant college and the newly approved experiment stations, which were affiliated with the land-grant college.

Nonetheless, the use of Hatch funds became a serious issue in Alabama. A branch agricultural experiment station, named the Canebrake Agricultural Experiment Station, had been established in west Alabama. This station was supported by state appropriations. In the late 1880s, with the availability of Hatch funds, an additional $2,000 of federal money was allocated for the Canebrake site. The precedent had been set. The state then created high schools to offer educational opportunities to students. These schools were dubbed "agricultural schools" and designated branch experiment stations in an apparent attempt to support them through the Hatch appropriation (Kerr, *History of the Alabama Agricultural Experiment Station*). Fortunately, the governor vetoed this proposal after the state legislature passed legislation to provide such funding.

The Alabama situation represented an extreme case of trying to use Hatch funds for things other than research. However, many states attempted to subvert the intended purpose of these funds. The problem of using research funds for other purposes exists to this day, albeit in a more subtle way. For example, large animals are used in teaching programs in animal science. In many states agricultural teaching budgets may not be sufficient to maintain these animal populations. Of course, experiment station animals are pressed into service as a result. This is a fact often overlooked by university administrators. The costs of teaching courses involving food animals such as hogs and cattle are far greater than those of teaching a biology course, where dead cats, frogs, pickled fish, or rats are used.

The importance of and the need for agricultural research were unquestioned. Yet, the compatibility of research and teaching gave rise to serious debate. Some held not only that teaching and research were compatible but also that synergism occurred when the two were carried out together. On the other hand, others argued that teaching and research were not compatible. In fact, even to this day teaching is often relegated to a lower status than research, although most forward-looking administrations see teaching, research, and extension in agriculture as equal callings. Fortunately, other than an occasional clash over resources, teaching and research are clearly a unified effort and combine effectively with the third component of this system, extension or outreach. Today, in most land-grant institutions many professors have joint appointments that provide them with an opportunity to engage in some combination of teaching, research, and extension. The percentage of assignment in each of these areas varies greatly and is usually determined by the individual's interests and capabilities and the needs within the systems.

When co-located with a land-grant university and a state agricultural experiment station, USDA-ARS scientists and forest research station scientists often have an opportunity to obtain adjunct appointments with the university. This greatly facilitates collaboration on research and enables ARS and forest service scientists to teach or direct graduate students and be a part of the academic process. Of course, a provision must be made to purchase the necessary time of the federal employee for state-supported work.

The framers of the Hatch Act and other related bills anticipated to a certain extent the potential conflict between the land-grant teaching and research programs. A wide range of agricultural leaders took the visionary step of creating an organization outside of government to bring together these related efforts. Such an organization, comprising those individuals who are working in the system, would be in a position to chart a course for the system and ensure its success. It would be based on leadership provided by those engaged in the effort. Thus, in 1887, the Association of American Agricultural Colleges and Experiment Stations was founded (Kerr, *Legacy*).[6]

During the past 125 years, this organization has grown far more complex, thus reflecting a much wider and more diverse membership. It has been particularly effective in providing a means of coordinating the efforts of various members of the association. One of its primary functions is to provide a unified voice for budget requests submitted to Congress for the support of research and educational programs.

The critical role agriculture played in the early history of the organization is sometimes overlooked today. When the leadership of the National Association of State Universities and Land-Grant Colleges invited the chair of the Board on Agriculture Assembly[7] to become a nonvoting member of its board of directors in 2007, some negative comments were made about why agriculture should have such a highly visible role in the organization. Nevertheless, this action reflected the pivotal role agriculture played in the formative years of the organization, as well as the importance of the agricultural enterprise today.

The language of the 1887 Hatch Act was unique. Once federal money was appropriated and allocated to the states, the federal government no longer had control, and federal administrators were dismayed by this arrangement. Consequently, the administration proposed an organization, called the Office of Experiment Stations, to help coordinate the agricultural research facilities. Is it any wonder that today the lay public or even a person coming into the system often fails to understand just how it is organized and how it works? Fortunately, during the early years a number of visionary leaders guided the development of the system. The teaching versus research debate and ultimate accountability came to a head with the 1899 ruling by the US attorney general that Hatch funds could not be used for teaching. Even though this ruling is in effect today, the highly diverse funding portfolio in most land-grant universities enables the administrative head of agricultural programs to move money in such a way that almost anything can be accomplished—and done so legally. For example, time and dollars are interchangeable, and assigning faculty different percentages of responsibility does free up resources for other areas.

Funding Issues Were Critical

A number of individuals possessed great vision and happened to be in key positions to contribute to the development of the agricultural research system we enjoy today. Wilbur O. Atwater was the director of the Connecticut Experiment Station and first director of the Office of Experiment Stations in the USDA. Before he became an administrator Atwater was an outstanding scientist with a distinguished record of accomplishments. While the need to solve practical problems was much of the impetus for initiating the agricultural research effort, the importance of basic research has long been recognized.

As the nation's first agricultural experiment station director, Atwater said in the Connecticut station's first annual report, "It has been felt from the first that more abstract scientific investigations would afford not only the proper but also the most widely and permanently useful work of an Agricultural Experiment Station" (Kerr, *Legacy*). Very true! His legacy was that experiment stations must focus on basic research to solve long-range problems.

Another of the early giants was Alfred Charles True, who guided the Office of Experiment Stations from 1893 to 1915 (True, *History of Agricultural Experimentation*). True followed Atwater's philosophy and perhaps as much as anyone set the tone and course for the agricultural experiment station system that exists to this day.

With the innovative funding mechanisms of the Hatch Act and the variations in philosophy of teaching and research, coordination was greatly needed. The Office of Experiment Stations and the Association of American Agricultural Colleges and Experiment Stations were helping, but still more was required. To address this situation, the Experiment Station Committee on Organization and Policy (ESCOP) was initiated in 1905. This was another effort by the experiment station directors to be proactive in addressing some of the conflicts within the system. The ESCOP is not to be confused with the Experiment Station Section of the Association of Public and Land-grant Universities (APLU) Board on Agricultural Assembly. Even some newly minted experiment station directors are unsure how each of these organizations really works and what their specific role is.

Several initial difficulties arose in getting the experiment station funding under way. The first payments to the states under the Hatch Act were delayed because of an error in the appropriation bill. A corrected appropriation act took care of this problem. In 1894, the secretary of agriculture denied funding to the experiment stations because he believed that the federal monies were not being properly accounted for. Though the funds were restored, a review system was put in place to shore up accountability. Finally, as stated earlier, in 1897 the US attorney general sought to clarify the unique funding mechanism and stated explicitly that experiment station funds could not be used for teaching.

Other legislation followed the Hatch Act in support of agricultural research. The Adams Act of 1906 provided for increased funding for research by the agricultural experiment stations established under the Hatch Act of 1887. Also, to ensure greater accountability of federal funds, the first research project system was implemented. The Purnell Act of 1925 provided for more complete endowment and maintenance of the agricultural experiment stations.

The Bankhead-Jones Act of 1935 supplied additional funds to the states for research in the agricultural experiment stations. Formula funds distribution, based in part on rural and farm populations, was implemented. While 60% of the funds were allocated to the states, the Department of Agriculture controlled 40%, which was used to create the USDA regional research laboratories, one to be located in each of the four major regions of the United States.

An amendment to the Bankhead-Jones Act in 1946 and the Agricultural Marketing Act of 1946 allocated further support for research and provided specifically for cooperative regional research (25%) and for marketing research (20%) by state agricultural experiment stations. As federal funding for the stations was institutionalized, some states took this as a signal to remove state funding for them. Since the station director has control of the Hatch funds, several university administrators chose to reallocate those monies to various other components of the university.

Regional research carried out at the state experiment stations went through several changes but proved to be a very creative and positive approach to enhancing cooperation among scientists and to addressing problems in agriculture. A "committee of nine" was created to review projects and proposals supported by the regional research effort.[8] Funds were to be allocated to individual projects and to each participant in the research effort. Though beautiful in concept, this arrangement provoked much bickering over funds.

Eventually, the regional research funds (RRF) were simply allocated to the states, who were then directed to account for their use on RRF-approved projects. Some states used the RRF funds as "carrots" to stimulate and encourage scientists to participate in research that would have a regional impact. Though this approach did have the effect of stimulating cooperation among several states on important researchable problems, it fell short of facilitating the extremely close cooperation that was originally envisioned.

However, a significant portion of Hatch money is still being allocated for multistate projects. Each station must track its expenditures against these funds to ensure they are spent on multistate research projects. An elaborate project-approval, participation, and reporting system is in place to facilitate the allocations and tracking. Experiment station directors in each of the regions work collaboratively to ensure the integrity of the multistate research function.

Other Significant Legislation Affecting Agricultural Research

Another milestone in the evolution of agricultural research was a bill enacted in August 1955. This Act of 1955 Consolidating the Hatch Act and Laws Supplementary Thereto states that "It is the policy of Congress to continue the agricultural research at state agricultural experiment stations, which has been encouraged and supported by the Hatch Act of 1887, the Adams Act of 1906, the Purnell Act of 1925, the Bankhead-Jones Act of 1935 and title I, section 9, to that Act as added by the Act of August 1946" (Kerr, *Legacy*). Clearly, this bill reflected Congress' intention to support agricultural research in the state agricultural experiment station system. Unfortunately, it did not spell out the extent of such support.

In 1962 the McIntire-Stennis Act was enacted. This legislation provided for management of and support for forestry research programs that were intended to work in association with each state's land-grant college and agricultural experiment station. However, provisions were made to accommodate some states whose forestry programs were located at other state-supported or private institutions. The forestry programs provided for under this legislation are not to be confused with forestry research programs that are part of the US Forest Service and therefore fall under the National Resources and the Environment (NRE) mission area of the USDA.

The 1977 Farm Bill included for the first time a specific title devoted to research. While the title refers to "research," it addresses teaching and extension (outreach) where appropriate. One section provided for support of agricultural and forestry extension at the 1890 institutions and Tuskegee Institute. Another section provided funding for agricultural research. Also included were provisions for a research grant program to make competitive grants for periods not to exceed five years. The bill also authorized the establishment of the Joint Council on Food and Agricultural Sciences and the National Agricultural Research and Extension Users Advisory Board. Both groups were to advise the agricultural research programs.

The research title of the Agriculture and Food Act of 1981 authorized research programs on rangeland and aquaculture. Sustainable agricultural research came with the Food Security Act of 1985, which provided for the establishment of controls on the development and use of biotechnology in agriculture.

National Research Initiative: Creating a Shared Leadership Vision

The Food, Agriculture, Conservation, and Trade Act of 1990 saw the expansion of competitive grants through the National Research Initiative (NRI) (Jordan et al., *Leadership in Agriculture*). The 1996 Farm Bill eliminated the Joint Council on Food and Agricultural Sciences and the National Agricultural Research, Extension, Education, and Economics Users Advisory Board. These boards were replaced by a thirty-member National Agricultural Research, Extension, Education, and Economics Advisory Board (NAREEE Advisory Board), which provides advice on agricultural research and related matters. Membership in the NAREEE Advisory Board is determined by nominations, and final review and appointment are overseen by the secretary of agriculture. Also, this farm bill established the Fund for Rural America to support research on rural development and community enhancement, international competitiveness, and environmental stewardship.

One of the more innovative approaches to strengthening the agricultural research effort in the country was the development of the National Research Initiative, mentioned earlier (Jordan et al., *Leadership in Agriculture*). The idea for the NRI was based on several premises, including (1) the great need for enhanced funding for agricultural research, (2) agriculture's importance to the economy, (3) efforts to enhance federal funding for the National Institutes of Health and the

National Science Foundation, (4) increased acceptance by some in the agricultural research community of the competitive grant-funding model, and (5) the emergence of biotechnology and its potential impact on agriculture.

The case for such a bold effort to support agricultural research was made by the Board on Agriculture and Natural Resources of the National Academy of Science, whose intent was to put agricultural science on a par with other major scientific efforts supported by the federal government. The idea appeared to be a slam dunk. Obviously there was a sound basis for establishing need (i.e., a bold $500 million per year request, strong support by almost everyone in the executive branch of government, including the agricultural research agencies, the USDA, and the US Office of Management and Budget, or OMB). A few key congressional supporters were committed as well. Support by the state experimental station directors was not unanimous, however. Most of those nonsupporters were convinced that such a new program based on the competitive grant model would drain funds from traditional formula funding.

All truly big ideas have many supporters, but usually one key leader emerges. For the NRI, it was Neville P. Clarke, who at the time was director of the Texas Agricultural Experiment Station. Clarke was an exceedingly eloquent, strong, and effective champion of the NRI. Indeed, he lived and breathed NRI from the beginning to the end and took a year's sabbatical leave from Texas A&M to advance the project. Also playing a key role in the development and marketing of the NRI were Charles Hess, USDA assistant secretary of science and education, and John Patrick Jordan, administrator of the Cooperative States Research Service (CSRS).

The goal of $500 million per year was never achieved; nevertheless, the NRI was the boldest effort in history to strengthen support for agricultural research. Today, the NRI has been incorporated into the Agriculture and Food Research Initiative (AFRI) of the National Institute of Food and Agriculture (NIFA), the replacement agency for the CSREES (Cooperative State Research, Education and Extension Service).

The 2002 Farm Bill, Title VII, provided for the continuation of existing programs but expressed the "sense of Congress" to double the funding for agricultural research during the next 5 years. It also provided enhancing funding for the Initiative for Future Agricultural and Food Systems. However, this particular program was viable for only a short period.

National Institute of Food and Agriculture Becomes a Reality

The Office of Experiment Stations was created in 1888 to oversee and coordinate the work of the state agriculture experiment stations. In the ensuing years it became the Cooperative State Research Service (CSRS) and later the Cooperative State Research, Education and Extension Service (CSREES). Then the 2008 Farm Bill ushered in yet another version of that organization, the National Institute of Food and Agriculture (as mentioned earlier), which comprises the following four program institutes, two administration offices, and several smaller staff units:

- Institute of Food Production and Sustainability
- Institute of Food Safety and Nutrition
- Institute of Youth, Family and Community
- Institute of Bioenergy, Climate and Environment
- Office of Grants and Financial Management
- Office of Information and Technology
- Center for International Programs, Budget Office, Civil Rights Office

The NIFA was created by Congress after much debate within the agricultural community. The Association of Public and Land-Grant Universities, a private group led by a distinguished citizen, and the USDA administration developed three somewhat different proposals for reorganizing CSREES. Congress took each one seriously. As created by that body, NIFA includes some of the provisions embedded in all three proposals.

The Agriculture and Food Research Initiative, which replaced the NRI, currently serves as a major funding initiative for research, education, and extension in a broad array of problem areas, including renewable fuels, energy, feed stocks, and energy efficiency. The 2008

Farm Bill also provides funding for research in organic production and for the Beginning Farmer and Rancher Development Program. It also states that the undersecretary for research, education, and economics shall be the USDA's chief scientist. The administration proposed changing the name of the research, education, and economics mission area to "Office of Science." Unfortunately, this was not included in the final farm bill legislation.

The foregoing section illustrates the dynamic nature of agricultural research in the United States. However, the many changes over the years reflect a great uncertainty as to how best to conduct these critical investigations. Their acknowledged success is more fully described in a later chapter. However, as indicated by the constant change, reorganization, and lack of adequate funding, Congress has failed to put into play a consistent and truly satisfactory model with dependable funding for agricultural research. One of the unique aspects of this enterprise is its grassroots nature, which has been evident since the beginning. It is demonstrated not only in the general need for research but, in many cases, also in the specific research needed.

The experience of developing agricultural research reflects the best of our system of governance. The people saw a need, articulated it, and insisted that the government respond in a positive manner. Some remain optimistic that the system can work in the future as it has in the past. Admittedly, almost everything has changed, with ever-increasing levels of complexity. Constantly increasing challenges will place greater demands on the agricultural enterprise and will demand a steady infusion of new information, knowledge, and technology that can be provided only through agricultural research. I am confident this is an achievable goal.

4
Changing Paradigms in Agriculture

And he gave it for his opinion, that whosoever could make two ears of corn or two blades of grass to grow upon a spot of ground where only one grew before, would deserve better of mankind, and do more essential service to his country, than the whole race of politicians put together.

—JONATHAN SWIFT

Progress in agriculture is characterized primarily by small, incremental improvements. In plant breeding modest increases in yields or slight resistance to a plant disease may lead to such gradual improvements. In soil fertility, simple advances in the placement of fertilizer or the timing of application could perhaps result in a slight increase in yield. Countless examples of such developments have collectively had a great impact on agricultural productivity.

There have also been major innovations that brought about new paradigms for agriculture. Sometimes these resulted from breakthroughs in research, while others occurred because of situation changes or developments in other areas. A number of them are chronicled in *A History of American Agriculture, 1607–2000*. Just a few of these inventions and developments collectively had a great impact on agriculture. In 1793 Eli Whitney invented the cotton gin, and the idea for a mechanical cotton harvester soon followed although it did not emerge as a successful tool until the mid-twentieth century. John Deere and a colleague started making and selling steel plows in 1837, some 40 years after the iron plow hit the market. Barbed wire was patented in 1874, making possible the end of unrestricted or open-range cattle grazing, which some states did not abandon until the mid-twentieth century. In 1955, the use of sterile screwworm flies made possible their eradication in the United States. These examples represent a small sample of the myriad developments that have contributed to agriculture's success in this country.

When considering agriculture as a continuum, with its beginning estimated at 10,000–12,000 years ago, a number of revolutionary developments have taken place. Though it might have taken many years to fully implement such advances, each one nonetheless brought about dramatic changes in agriculture. These dramatic changes I refer to as "new paradigms."

Simple Hand Tool

Probably the first real innovation in agriculture was the invention, development, and use of some form of hand tools, which were developed by early human beings. The first agricultural hand tool was something like a hoe, which probably started as a stick. The stick was sharpened; then a sharp object such as a shell or flat rock was fastened to it. The use of hand tools, while simple and rudimentary, had a major impact. The hand tool use period probably lasted several thousand years, and they are still in use today in some form. Some version of the hoe is still used on commercial farms and is often the tool of choice of the home gardener today.

Domestication of Animals

Animals have become so intertwined in our culture that we often fail to realize what an impact they have had on agricultural productivity and, indeed, civilization. Considering the countless species of animals on this planet, only a few have been domesticated. First, only terrestrial mammals have been domesticated, and only five species have become widespread and important throughout the world: sheep, goats, cows, pigs, and horses (Diamond, *Guns, Germs, and Steel*). Jared Diamond notes an additional nine species: the Arabian camel, Bactrian camel, llama/alpaca (distinct breeds of the same ancestral species), donkey, reindeer, water buffalo, yak, banteng, and gaur are important in limited areas of the world. Diamond notes six criteria for domestication: flexible diet, reasonably fast growth rate, ability to breed in captivity, pleasant disposition, good temperament,

and modifiable social hierarchy(ibid.). These are demanding requirements for wild animals.

Early human beings evolved along with many other mammals.
Probably the first animal domesticated was the dog some 12,000 years
ago. The dog was, of course, the first companion animal. This relationship or bond between people and dogs continues. Some would argue
that cats have yet to be domesticated; they have been tamed perhaps
but are not fully domesticated.

Domestication of sheep, goats, pigs, and cattle occurred 8,000 to
10,000 years ago. These animals provided a consistent and more reliable source of food, particularly protein, than the kills made by hunters. Growing up in a rural environment, I personally had opportunities to sample opossum, raccoon, rabbit, duck, rattlesnake, dove,
squirrels, and a few other wild creatures. There is no question that
pigs and cattle are an improvement as meat animals. Our ancestors
did well in deciding which animals to domesticate.

Domestication of cattle deserves special note because of the importance of this particular species to human beings. Cattle met the
aforementioned criteria for domestication admirably. In addition,
they provided milk and related products. Cattle survived by living on
most things not suitable for human food and were able to do so for
long periods, thereby expanding the harvest season. More important,
they were adapted as beasts of burden and as a power source that
augmented man's energy. Other large animals such as the American
bison have been widely used as well. Even cattle that are "finished" on
grain spend most of their life living on grasses, which are not eaten
by human beings.

Today's cattle were derived from an animal now extinct, the aurochs,
or European wild ox (Bradley, "Genetic Hoof Prints"; Bradley et al.,
"Genetics and Domestic Cattle Origins"; Götherström et al., "Cattle
Domestication"). Daniel G. Bradley and his colleagues describe the
Bos taurus and *Bos indicus* as well as a third type of cattle in Africa that
is somewhat between *B. taurus* and *B. indicus* (Bradley et al., "Genetics and Domestic Cattle Origins"). Various species of the genus *Bos*
were first domesticated about 10,000 years ago. First was *Bos taurus*
and then later, around 7,000 BC, *Bos indicus*. The latter are humped
cattle, characterized today by the Brahma and a few other breeds. My

guess is that very soon after domestication, early human beings hit on the idea of having these animals serve as beasts of burden that would pull their crude but steadily improving simple implements, tools, and wagons. One can only imagine how the early years of animal power looked and evolved.

Animal power is still much used in many parts of the world today. Animals powered much of American agriculture until the invention of the internal combustion engine and subsequent developments of the gasoline- or diesel-powered tractor. Prior to the internal combustion engine tractor, the steam engine, developed around the end of the seventeenth and the early eighteenth centuries, was adopted as a power source for agriculture. For several reasons, not the least of which was cost, steam never evolved as a major source of power for this purpose.

Another noteworthy domestication was that of the horse and the donkey, which occurred about 6,000 years ago. Upon domestication of the horse, people quickly recognized its potential transportation advantages. Of course, the horse was a natural for warfare, where it was widely used until the early days of World War II, when the tank rendered the horse cavalry obsolete. In fact, for several thousand years there was no better form of transportation than the horse until the invention of the steam locomotive.

While cows and horses often served as a power source, other close relatives made quite significant contributions as well. In India and much of Asia, the water buffalo was the choice beast of burden. The hybrid offspring of a male donkey and a female horse is a mule. Both the water buffalo and mules are still widely used in much of the world today. Throughout much of the United States, horses were used extensively, while mules were most common in the South.

A number of other animal species have also been domesticated, including the silk moth, camel, chickens, honeybee, rabbits, and turkeys. Probably the most unique of these is the wasp, which has been trained to detect various explosives (Lewis and Tumlinson, "Host Detection"; Rains et al., "Behavioral Monitoring").

One thing that is resoundingly clear is that animals have been a part of our survival and our well-being from the very beginning of our civilization. Animals comprise a major component of our diet and are used in many other ways to support our well-being.

Domestication of Plants

The development of some plants into crop plants is equal in importance to the domestication of animals. As described earlier in this book, early human beings found that certain plants had qualities that in some way met their desires and needs. Obviously, they favored the plants that met those simple criteria. This was the first step in the domestication process, which was a crucial part of the development of agriculture and enabled a more sedentary lifestyle. The development of food production (agriculture) emerged in different parts of the planet. Diamond mentioned five such areas, including Southwest Asia (generally thought of as the Fertile Crescent), China, Mesoamerica, the Andes region of South America, and the eastern United States (Diamond, *Guns, Germs, and Steel*). He further points out that the earliest dates of domestication varied widely, from about 8500 BC in Southwest Asia to about 2500 BC in the eastern United States.

The domestication of plants and the emergence of agriculture were closely related to the characteristics of native species. The indigenous species of grasses that gave rise to modern wheat and barley already possessed some desirable traits, such as large seed size and nutritional composition.

Domestication of plants was probably the beginning of modern agriculture, yet it remains a work in progress. Much active research focuses on plant exploration in an effort to identify indigenous, wild relatives of crop plants that have some unique genetic properties that can be incorporated into currently used crop plants. Today, virtually all food crop plants have been genetically modified either by selection using classical genetics or biotechnology techniques.

Plants are unique and occupy an important niche in agriculture. By utilizing chemical nutrients in the presence of energy from the sun, they can produce their own food. Consequently, plants provide nourishment for people and animals.

There are four major groups of plants—mosses and liverworts, ferns, conifers, and flowering plants. Various authors estimate the existence of 300,000–400,000 species of plants. Of these, only a few thousand can be eaten by people, and just a few hundred have been domesticated. Only a few dozen species of plants contribute to the food supply on this planet. In fact, the cereals (e.g., wheat, corn, rice, barley, sorghum) account for about half of the calories consumed by people. Diamond (*Guns, Germs, and Steel*) notes that 80% of the tonnage of all crops is accounted for by about a dozen plant species, including cereals; pulses; roots or tubers (e.g., potatoes, manioc, sweet potatoes); sugar sources (e.g., sugarcane, sugar beets; and fruits (e.g., bananas).

Food Preservation

An important development in the evolution of agriculture was food preservation. This was especially important when quantities of food were abundant, ensuring there would be something to eat when food became scarce, as it often did. Concern became even greater as cooking and preparing food for consumption became more prevalent. The first means of saving food for difficult times occurred as seed plants evolved. Seeds provided an excellent means of preservation. A plant could produce a food that could be eaten immediately or be saved and brought out at a later date and used. Seeds still comprise a significant component of our diet today. However, before various means of preservation were developed, they meant that food could be preserved in times of plenty and then produced (grown) when it was not readily available. In fact, some attribute the success of our early civilization to the development of crops that produced abundant seed. Supporting this idea is the knowledge that our civilization initially developed in regions of the world where seed plants (e.g., wheat, barley, other grains) evolved (Sinclair and Sinclair, *Bread, Beer and Seeds of Change*).

In addition to natural preservation approaches such as seed, other means have been developed over time. It is particularly noteworthy that methods of food preservation dating back to the early days of our civilization are still relevant and used in some fashion today. Since our foods generally involve water systems, preservation by early people consisted of removing or tying up water in food to prevent spoilage.

Preindustrial Era

Smoking. After the discovery of fire, early human beings learned that meat exposed to smoke would still be suitable for eating longer than the fresh kill. We now know that some compounds in smoke help prevent growth of certain microorganisms that cause spoilage. Though smoking no longer serves as a survival means for people, it is still widely used to improve the flavor of certain foods, particularly cheeses, nuts, and meats such as beef, pork, lamb, poultry, wild game, and fish.

Drying. One of the oldest ways to preserve meat and vegetables is simply to remove the water, thereby limiting or preventing growth of microorganisms. In all likelihood drying food products dates back a few thousand years and is still widely used today, making possible such foods as raisins, prunes, and sundried tomatoes. Another delectable food is prosciutto, which is ham that has been cured by drying. However, salt and other chemicals are often used in processing to attain the desired taste. Another dried food that has become popular today is beef jerky, which is widely sold in convenience stores.

Salting. Another of the earliest approaches to preserving meat, fish, and vegetables was the use of salt (NaCl). Indeed, salt played a key role in our early civilization. Though the exact time of discovery of salt is not known, it is clear that it has long been recognized for its unique properties. Securing salt for use in preserving meat was only the first step. Salting, along with drying and smoking, probably were the best means of preserving food that could be consumed during the long winters and droughts, while traveling, or during spells of poor hunting.

Indeed, salt was so important it was often interchangeable with money. Today we take salt for granted; however, until about 100 years ago, salt greatly influenced the status of civilization. It is quite important to an agricultural society for the following reasons. Unlike hunters, who fulfill their salt needs from wild game, humans and livestock in an agricultural society must consume salt. Also, salt is crucial for food preservation, which is critical in agricultural trade. Preservation of food by salting is still commonly used today, although other means of preservation are more prevalent.

Fermentation. Use of fermentation in preserving food can be traced back to the Egyptians (2800 BC). The end products of fermentation are alcohols and acids. Highly important foods such as cheese, alcoholic beverages, and sauerkraut (an acidic food) are the resulting products of fermentation. Fermentation is a wonderful process that enables the change of bland foods into flavorful foods. For example, grapes are suitable for eating for only a few days after harvesting. However, once fermentation turns their juice to wine, they can be enjoyed all winter. Grains (e.g., barley, wheat, rye) make wonderful breads, but after fermentation turns them into beer, their usefulness is greatly expanded.

Pickling. Fresh-packed pickles, as well as fruit pickles and relishes, can be made by adding vinegar or other acids to them. These foods can be produced in 24 hours. These products differ from "fermented pickles," which take weeks to mature. Pickling is widely used today with eggs, peppers, herring, cucumbers, and other vegetables.

Lime Curing. To make dried fish more palatable, Scandinavian countries during the Middle Ages cured them with lime to facilitate water uptake and swelling of the tissue. This age-old custom is still practiced to preserve cod, pike, and ling (a member of the cod family) to produce a dish called lutefisk. It is also employed in the production of olives and hominy (corn).

Burial in the Ground. This process has long been used to store root crops such as sweet potatoes and onions. In preservation by burial, some type of mulch must keep the food product from coming into contact with the soil. Dark conditions and a small amount of ventilation is needed for root vegetables to "breathe."

Industrial Era

Canning. Canning or bottling are excellent ways to preserve food. Acidic foods such as most tomato varieties and fruits are sealed in a jar or can and boiled in a water bath. Nonacid foods such as meats and vegetables are sealed in a jar or can and "retorted" at high pressure (240–260° F) to ensure their shelf life and safety. Canning and bottling are widely used means of preservation of food today by the commercial sector and in the home.

Concentration. Some meats and fruits can be preserved by jellying. Meats such as eels provide a gel when cooked. Jellying of marmalade, preserves, jellies, and spreads requires cooking fruit with sugar or corn syrup, pectin, and acid to 65–68% solids.

Freezing. The invention of refrigeration ushered in an opportunity to preserve certain foods. Freezing is the process of lowering the temperature of a food below 32° F so that the growth of microorganisms is retarded. Slow freezing (such as placing food in the freezer compartment of a refrigerator) will kill some microbes but not all. Those that produce spores are usually not injured. Freezing is probably the most common means of preserving food today.

Modified Atmosphere. A relatively recent development with regard to certain crops is the identification of the specific atmosphere that will facilitate preservation. For example, reducing oxygen while increasing carbon dioxide helps preserve some salad crops, at least for short periods. Controlled atmospheric storage of crops such as apples and Vidalia onions has had a dramatic impact on the entire apple and Vidalia onion industry by extending the period of availability. (Chapter 7 explains this in more detail.)

Irradiation. One of the more recent approaches to preserving food is to expose it to ionizing radiation, which can be provided by several sources (e.g., cobalt-60, cesium-137). Though irradiated food does not become radioactive and poses no threat to human health, such food preservation approaches have not been totally embraced by the public at large. However, there appears to be some increasing acceptance of irradiation as a means of preserving food. It is important to note that many preservation techniques employ rather basic scientific principles and skills. However, some recent developments such as irradiation of foods are many steps ahead of the appreciation of science by the general public. The science of food preservation continues evolving to this day.

Mechanical Devices and Tools

Farmers at the time of the American Revolution had tools for farming that were little better than those of the early Egyptians. In fact, the ancient Egyptian farmer would not be lost on the typical farm in England or America in 1776.

Technical Inventions

All of this changed dramatically with time and innovation. The Industrial Revolution covered much of the eighteenth and nineteenth centuries. Hardly any aspect of human endeavor escaped the changes it brought about. Manufacturing, transportation, and, of course, agriculture were dramatically affected. Development of various mechanical devices and tools ushered in a new agricultural era. Many of these were simple, yet led to greater efficiency. One of the earliest such tools was the sickle. Another was the plow, the forerunner of which was an ard (Sinclair and Sinclair, *Bread, Beer and Seeds of Change*), a simple device pulled first by humans and later by animals. The first plow was probably a rather crude device. There are so many incremental developments of the plow that it is difficult to assign dates or credit to specific individuals.

The inventor of the first plow or its forerunner, which was likely of wood, is not known. In all likelihood the idea of some crude type of plow dates back several hundred years. Some credit the Chinese with employing a flat piece of iron or wooden blade to facilitate plowing. About the best that we can do is to identify the general period in which the modern plow was developed and some of the people who contributed to its development.

Toward the waning days of the eighteenth century the idea of some type of a metal plow emerged in both England and the United States. In 1797 a patent was issued in the United States for a cast iron plow. Other improvements followed with patents in 1807, 1814, and 1819. As one would expect with such rapid development, patent rights stirred up considerable challenges. These efforts culminated in the design of a cast steel plow by John Deere in 1837. This design proved to be the tool needed to break up the heavy, fertile prairie soils of the Midwest. That Deere's invention was successful is further supported by the survival of the John Deere Company, which flourishes to this day.

Another improvement to the plow is the development of a harder, smoother, glossy surface that facilitates plowing in heavy clay soils. The disk plow was developed in the mid-nineteenth century. It was soon realized that two or more plows could be fastened together. As with most innovations in agriculture, introduction of the cast iron and later the steel plow was not without controversy. Some naysayers maintained that these plows would "poison the soil," resulting in a deleterious effect on agricultural production. Improvements and innovations of the plow continue even today. One of the most recent is the development of the switch plow, which allows the operator to make an adjustment such that the soil can be turned either to the right or to the left. This enables the operator to plow in all directions.

While the plow was evolving, other innovations were emerging. Development of the cotton gin, a machine that could quickly separate the cottonseed from the cotton fiber, revolutionized the cotton industry in the United States. Eli Whitney is usually given full credit for inventing the cotton gin in 1793. However, because of a failure to effectively protect his patent, he failed to profit from his invention.

During this period developments occurred in the harvesting of crops. The cradle scythe made its appearance in the latter days of the eighteenth century. Next came the mechanical reaper, patented by Cyrus McCormick in 1834. This animal-powered machine replaced the manual cutting of small grains with sickles and scythes. Another innovation was the threshing machine, which mechanically separated the grain from straw and chaff, substantially reducing the drudgery of flailing grain by hand. Though the concept of a thresher emerged early in the eighteenth century, recognition for the invention goes to Andrew Meikle, a Scottish mechanical engineer, in 1784. As with most such innovations, other inventors made substantial improvements or offered alternative approaches. More practical threshing machines were patented in the United States in 1837, and a machine was developed in Australia in 1843.

Corn pickers were introduced in the mid-nineteenth century. While these early devices simply removed the ears of corn for further processing, today almost all corn is harvested such that only grain is removed from the field.

Cotton harvesting has taken two approaches: stripper harvesting, which removes the entire boll for further processing, and spindle pickers, which remove only the seed and fiber. Though the concept of cotton harvesting has been around for many years, it was not until the mid-twentieth century that machine harvesting of cotton was widespread. With conventional cotton pickers, compacting cotton into modules for ease of transporting to the gin is standard. In recent years, these two processes have been combined by using round bales of cotton. Early in the twenty-first century patents were issued for devices that use high-velocity air to remove the cotton fiber from the bolls. The success of this approach remains to be validated.

Peanut harvesting has undergone a radical evolution—all in the twentieth century. Until the mid-twentieth century peanut harvesting consisted of several steps: First, the peanuts were extracted from the ground either by hand or plowing, using a single moldboard plow minus its wing. They were then collected and stacked on temporary poles throughout the field. After curing (drying), the peanuts were removed from the vine with a stationary thresher in the field. Today peanuts are dug with a digger equipped with an inverter-shaker, thus exposing the peanuts to the sun for drying in the field. When the peanuts are sufficiently dry, they are combined (threshed) by large self-propelled combines that accumulate the peanuts for depositing in large trailers. Only a modest amount of manual labor is required.

Improvement of the harvesting of crops continues to evolve, leading to greater efficiency and convenience; however, challenges to the mechanical harvesting of some crops remain, particularly horticultural crops such as squash, peppers, eggplants, okra, cucumbers, and strawberries, which are still harvested primarily by hand.

The foregoing innovations in agriculture were not heavily dependent upon science. However, they required a remarkable level of engineering creativity. But, beginning with the nineteenth century, new developments in agriculture were essentially based on science, which depended on research.

Fertilization

One of the first of the science-based innovations in agriculture was the concept of nutrient modification, commonly referred to as fertilization. The concept of replenishing (i.e., returning nutrients to) the soil was a major agricultural development. Our civilization is totally dependent upon the productivity of that thin, upper layer of material called topsoil. While it is comforting that soil does not necessarily wear out or is used up, it is equally true that specific elements in the soil must be replenished or replaced in order to keep the soil productive. This is because specific nutrients are taken up by the plant and removed from the land at harvest. Some are also lost through leaching and erosion by water or wind. Regardless of how they are used or lost, in order to ensure continued productivity, these nutrients must be replaced on a continuing basis.

In his book *Commercial Fertilizers,* Gilbeart Collings offers several suggestions as to how fertilization began. River bottoms were probably cultivated in early civilizations because they were more fertile. Runoff from the hillsides moved fertility to these lower levels, which were more productive than the uplands. Most archaeological investigations show that the culture of some crops started around 10,000 to 12,000 years ago. Early farmers probably observed that animal manure enhanced growth. From this, it would be a small step for early farmers to collect animal manure to put around desirable plants. Throughout early history, many other products such as bones, wood ashes, guano, chalk, and fish were collected to use as fertilizer. Though these led to enhanced productivity, they alone hardly brought about a new paradigm for agriculture. However, in the mid-nineteenth century a number of discoveries ushered in the era of commercial fertilizer.

In 1665 it was observed that that saltpeter (KNO_3) would increase crop yield. At the turn of the century another observation indicated that ingredients of ash were essential for plant growth. It was Justus von Liebig, a noted German chemist, who showed that potassium and phosphorus were necessary for plant growth (Liebig, *Organic Chemistry*). Saltpeter was found in China and India but was used primarily in gunpowder.

Knowledge of the need for plant nutrients for suitable agricultural productivity grew along with the identification of the specific chemicals required for growth. This created an opportunity to fulfill the need for fertilizer. Early fertilizer materials included ground bone, Peruvian guano, and, later, Chilean sodium nitrate and potassium salts. These materials covered the big three (nitrogen [N], phosphorus [P], and potassium [K], or NPK) plant nutrients. The discovery of the role of plant nutrients in crop productivity and the importance of fertilizer in addressing this need had a far-reaching impact on agriculture.

Justus von Liebig demonstrated that the fertilizing value of bone meal could be enhanced by treatment with hydrochloric or sulfuric acid (HCl or H_2SO_4, respectively) (Liebig, *Organic Chemistry*). In 1842 John Bennett Lawes patented a process to treat rock phosphate with sulfuric acid (Lawes et al., "On the Sources of the Nitrogen of Vegetation"). The result was superphosphate, a mixture of calcium phosphate and calcium sulfate, which is more soluble and therefore more desirable as a source of phosphorus fertilizer (Sisler et al., *College Chemistry*).

To evaluate the effect of these materials on plant growth, investigations were initiated at the Rothamstead Experimental Station. The importance of long-range fertility studies that began at Rothamstead have continued to this day with only a few modest changes (*Rothamsted Experimental Station*). The value of such experiments was recognized by researchers in the United States. The Morrow plots on the campus of the University of Illinois are the second oldest fertility-crop rotation plots in the world. They were established in 1876, only 33 years after similar plots were established at Rothamstead in England. After 138 years, these plots have yielded valuable data on the effects of crop rotation, natural soil nutrient depletion, and various fertilizer materials on crop yields. The "old rotation" at Auburn University was one of the first experiments to demonstrate and document the value of rotating cotton and legumes.

The Magruder plots established in 1891 at Oklahoma State University as a winter wheat fertility study have yielded volumes of useful information (Mullen et al., "Magruder Plots"). The experiment was initially established to evaluate the use of manure on wheat yield; however, later treatments were altered to evaluate inorganic sources of N, P, and K. Among the findings in this long-term fertility experiment was that it is critically important to recognize a soil's nutrient-supplying capability. The first limiting nutrient was phosphorus, followed by nitrogen and potassium.

Of special interest is the Sanborn field experiment, which was established in 1888 on the campus of the University of Missouri (Miles and Brown, "Sanborn Field Experiment"). This was the first facility in the United States designed to measure soil erosion under various cropping systems. Results from these plots contributed to the establishment of soil conservation policy in the United States. The fungus *Streptomyces aureofaciens* was isolated from one of the Sanborn field plots in 1945. This provided the original source of the antibiotic Aureomycin. Additional long-term fertility plots have been established at other agricultural research institutions. The creation of these first experimental stations adds to the significance of the fertilization paradigm. The concept of plant nutrition and the role of fertilization were new steps in agricultural evolution. It is a bit difficult to realize that this whole concept is only a century and a half old. While clearly this development played a key role in establishing experimental research, particularly at the Rothamsted Experimental Station, questions regarding the quality of fertilizer materials served as a great stimulus for setting up experiment stations in the United States.

Though there were multiple factors in the development of such stations in the United States, much of the impetus for creating them in many states prior to the Hatch Act of 1887 was the need for help in determining the quality of fertilizers that were offered for sale to farmers. Apparently some unscrupulous businesspeople were selling fertilizer that was little more than sand. Of course, without chemical analysis, it was not possible to know this. Exploiting the concept of fertilization required much basic science and dedicated research. While much has been learned about fertilizer and plant nutrition in the past 150 years, soil chemistry and soil fertility remain major research thrusts today. In fact, if I were to start an experiment station today, and I could employ only one person, that person would be a soil chemist. The new developments in genetics of crop plants and the continued search for more effective fertilizer materials and formulations seem to never end.

Fertilizer use helps stave off famine and lift people from poverty. It helps provide a stable foundation for civilization itself. (Hall and Nowels, Fertilizer 101)

In considerations of agricultural sustainability, one of the prime factors is sustainability of plant nutrients. Nitrogen, the first of the three major components of commercial fertilizers, was probably the first material to be sold as a commercial fertilizer. Though deposits of Peruvian guano were discovered early in the nineteenth century, they were not fully exploited as a fertilizer material until later in the century. The Peruvian guano lasted for much of the century.

Peruvian guano fulfilled two major functions. First, it was a great source of nitrogen fertilizer, and, second, it gave credence to the power of the concept of fertilization. The next step in the nitrogen story is the utilization of natural deposits of nitrate of soda in Peru and Chile. With the depletion of Peruvian guano and Chilean sodium nitrate, the primary nitrogen sources at the end of the nineteenth century were ammonium sulfate and sodium nitrate (Lawes et al., "On the Sources of the Nitrogen"). Some use was made of organic materials (e.g., cottonseed meal, blood, fish scraps, other animal by-products) as a nitrogen source. Today, much of our nitrogen is formulated from ammonia, which is derived from natural gas and atmospheric nitrogen through the Haber-Bosch process.

The second of the big three plant nutrients is phosphorus. Sources of phosphoric acid are basic slag, precipitated phosphate, and superphosphate. Phosphorus became popular as a fertilizer in Europe, particularly after von Liebig explained how to make superphosphate by treating bone meal with hydrochloric or sulfuric acid. At current use rates, optimistic projection shows the United States with about a 110-year supply, while the world's known reserves will be expected to last more than 200 years (United Nations, Population Division). The last of the big three nutrients is potash. Potash is supplied by chemicals such as kamite, manure salts, potassium chloride, potassium sulfate, and sulfate of potash magnesia. Much of the potash used in fertilizer in the United States comes from Canada. Sale of complete mixed fertilizer did not grow rapidly until the first quarter of the twentieth century. In the United States, use of commercial fertilizers increased rapidly between 1895 and 1945.

It is interesting to note that the supply of two of the big three fertilizer materials, nitrogen and phosphorus, is of great concern. Nitrogen is derived from air and combines with H_2 from natural gas in the Haber-Bosch process. Natural gas is a fossil energy source. Also, there is a limited supply of phosphorus. At present, some of the nitrogen fertilizer and more than 80% of the potassium fertilizer we use in the United States are imported. This is about to change with recent developments in natural gas and fracking technologies combined with renewable energy dynamics. The introduction of fertilizer into the crop-production model brought about a major paradigm change in agriculture. Lack of or a substantial increase in the price of any one of the big three plant nutrients, NPK, would bring about another major change—in the wrong direction.

This is a basic problem, to feed 6.6 billion people. Without fertilizer, forget it. The game is over.
—NORMAN BORLAUG

Consumption of commercial fertilizer in the United States was only slightly more than a million tons in 1880 and stayed below 10 million tons until 1943 (Collings, *Commercial Fertilizers*). During the past century, consumption has steadily increased (ibid.; Slater and Kirby, *Commercial Fertilizers 2010*). However, it appears that a plateau was reached in the first decade of the twenty-first century. Consumption (at 10-year intervals) is illustrated in figure 1. Even though fertilization research is one of the oldest areas of agricultural testing, many challenges and opportunities in the area of plant nutrition remain.

Soil analysis has become routine in crop production. However, simply testing for plant nutrients in the soil is of modest value. In order to achieve the maximum value from soil testing, it should be coupled with soil test calibration. This is a slow laborious process that is greatly influenced by soil type, crop cultivars, and other factors. Clearly for the most efficient production of agricultural crops, testing must be coupled with soil test calibration. New infrared scanning technologies may make soil analysis significantly faster.

Mechanical Power

Invention of hand tools, then development of animal-powered implements with ever-increasing sophistication ushered in a new era. Animal power in lieu of human labor was a revolutionary step forward. Even more dramatic was the introduction of mechanically powered implements, which changed agriculture forever.

Growing up on a farm in the southern United States that was powered by mules, I personally experienced the transition from animal power to mechanical power. There was unfettered euphoria when

Figure 1. Fertilizer consumption in the United States, 1880–2010 (ten-year intervals).

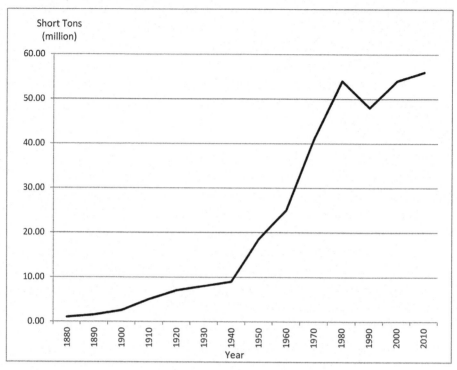

Sources: Collings, *Commercial Fertilizers*; Slater and Kirby, *Commercial Fertilizers*.

that first green tractor started plowing the fields on our farm. The transition occurred essentially overnight. Though some new implements were purchased with the new tractor, most of the mule-drawn tools were altered so they could be powered by the tractor. The first change that accompanied the entrance of the tractor was the comfort. You could do all the plowing, disking, planting, cultivating, and harvesting without walking. Just to plow a single 40-acre field one time with a one-horse plow required the farmer to walk 660 miles. The average farmer would walk several thousand miles to make a crop. The tractor could plow the biggest, heaviest infestation of weeds that sometimes stopped the mule-drawn plow. One person could accomplish much more with a tractor than with a mule. In addition, the tractor required no fuel when at rest. The mules and horses required feed 365 days of the year whether working or not. Feed for the mules or horses actually required a significant portion of the annual harvest.

The situation just described was not where mechanical power started on the farm, however. This was almost an idealized situation. The invention of the steam engine occurred over a period of many years, long before the internal combustion engine. Development of the steam engine involved the effort and contribution of many people. Ideas and concepts for a steam engine date back to the ancient Greeks. Even Leonardo da Vinci envisioned the steam engine. However, it was Thomas Newcomen who substantially improved a steam-powered machine to pump water. Later, James Watt made significant improvements on Newcomen's machine and was awarded an exclusive patent and the rights for his engine in 1869. It was during this period that the steam engine was adopted as a mechanical source of power—a steam traction engine or, in today's jargon, a tractor. Development of things such as farm tractors was delayed in the United States during this period because the country had more pressing matters to attend to—the US Civil War.

The steam tractor was a mechanical monster weighing up to 15 tons, but it was highly effective in pulling plows, disks, and so on and was especially helpful when using a belt to power threshers and other stationary machinery. Development of a more effective steering mechanism and a clutch made the steam tractor a bit more practical. However, as one would expect, the size, cost, and lack of versatility of the steam tractor prevented widespread use among moderate-sized and small-scale farmers. Also, one other drawback was that steam engines sometimes exploded.

What steam engines really did was to initiate the transition from animal power to mechanical power. Growing up on a two-mule farm in the United States in the 1940s and 1950s, even a small boy clearly recognized that the future of agriculture depended upon mechanical power. One of the last commercial uses of steam tractors that I am aware of was in southern Ontario, Canada, in the 1950s. In the flue-cured tobacco region of southern Ontario, farmers often experienced an early frost toward the end of the harvest season. Consequently, as soon as a barn of tobacco was cured (dried), the farmer wanted to remove the tobacco so the barn could immediately be used again. Normally, the farmer would wait a couple of days until the tobacco absorbed moisture from the atmosphere and became "in case" (i.e., soft and pliable). With live steam provided by a steam engine, this process could be readily accomplished almost immediately. Today steam tractors are a novelty seen only in state fairs or antique tractor shows, usually sponsored by special interest groups. The Agricultural Engineering Club at Auburn University has, as a student project, a steam tractor its members maintain and operate on special occasions.

The development of the internal combustion engine (ICE) is one of the major inventions of our civilization (*History of the Automobile*). Many scientists and engineers contributed to the concept, design, and development of the engine even prior to the modern era. A number of developments occurred during the seventeenth, eighteenth, and first half of the nineteenth centuries in development of the ICE. An internal combustion engine that used a hydrogen and oxygen mixture and was ignited by an electric spark was built by François Isaac de Rivaz in 1807.

The last half of the nineteenth century saw the ICE develop into the practical power source it has become today. Many individuals who contributed significantly to the developments of the ICE (e.g., Nikolaus Otto, Karl Benz, Gottlieb Daimler, Rudolf Diesel) are still well known today (*History of the Automobile*). Diesel unveiled his diesel engine in 1900, but this is hardly the end of the story. Further developments of the ICE continue to this day.

In view of the acceptance of the concept of mechanical power as illustrated by the steam tractor and the obvious versatility of the ICE, it is a small step to incorporate the ICE into many activities on the typical farm (e.g., pumping water, grinding feed). Of course, the next step is to go from a gasoline engine to a gasoline traction engine or tractor. It is interesting to note that the early gasoline tractors were similar to the steam tractors in that they were large, heavy, and expensive. It was not long before engineers realized these machines could be drastically scaled down. Early versions of the gasoline-powered tractor included massive steel wheels. A breakthrough came when research at the Ohio Agricultural Experiment Station showed that pneumatic tires could replace the steel wheels.

The transition from human labor to animal power was a significant development and a genuine paradigm change. However, the transition from animal power to mechanical power was earth shattering. While much of the United States experienced this transition in the 1920s and 1930s, it was just beginning to make inroads in the South when World War II erupted. Progress was instantly halted, and almost all tractor manufacturing in the United States ceased. Instead of building farm tractors, such efforts were directed to building tanks, trucks, and other heavy equipment for the war effort.

Agriculture was an integral part of the war effort as farmers were encouraged to maximize their production and to grow some unfamiliar crops (e.g., hemp) that contributed to the war effort. Consequently, a great pent-up demand for gasoline-powered tractors developed. Immediately after the war, American industry swung into action, producing farm tractors by the thousands. The transition to a mechanical power paradigm in agriculture became a reality.

Classical Genetics

Gregor Mendel, an Austrian monk who studied the inheritance of various traits of peas, ushered in the science of genetics. His work, between 1856–1865, allowed him to understand both segregation of genetic traits and independent assortment of traits. The result was Mendel's laws of inheritance. Prior to Mendel's research, plant improvement consisted only of phenotypic selection of the most desirable plants. This was not greatly unlike the approach employed by early human beings at the beginning of agriculture. With an understanding of Mendel's laws of inheritance, classical genetics enabled great strides in crop and animal improvement.

Scientists for the first time could systematically incorporate desirable genetic traits from one plant or animal into new progeny. This paved the way for dramatic improvements in agricultural productivity. Classical breeding involves interbreeding (crossing) similar but sometimes distantly related individuals to produce offspring that incorporates some particularly desirable trait. The intent is to bring new genes from distant relatives or unrelated individuals that possess certain unique traits or properties. For example, major crops such as soybeans can sometimes be improved by crossbreeding with similar species of soybeans collected from indigenous populations. This is particularly relevant for nematode and disease resistance by increasing yield potential, developing tolerance to adverse growing conditions, and so on.

Improving crops through selection is almost painfully slow, yet it has led to significant improvements. On the other hand, employing Mendel's laws of inheritance was also a slow process. However, it did lead to some measure of expected, desirable outcomes. More recently the classical plant breeder also employed other techniques to incorporate different genetic characteristics into new progeny. These included embryo rescue and the use of radiation. Such techniques were guaranteed to produce different traits in plants but not necessarily more desirable ones. Often the results were hit or miss. Classical genetics laid the groundwork for another new paradigm in agriculture—hybridization (a cross is a hybrid).

Hybrid Vigor

As early humans identified those particular plants that had desirable traits, they likely also began to favor those plants instead of others. From there it was a small step to begin selecting the best of those desirable plants. Even though selection is an inexact manner of plant improvement, it made great progress possible simply because it took advantage of the naturally occurring genetic variability of plants.

The process of hybridization enables the assimilation (cross) of the best genes of two (or more) genetic lines. Although selection and hybridization do not create new genes, they can take advantage of the best combination of genes already present in the parents. Of course, modern plant breeders are constantly in search of new genes that can be incorporated into existing germplasm. This process has enabled great strides in crop improvement.

Late in the nineteenth century it was noted that crossing different inbred lines often produced progeny that had certain highly desirable characteristics. Corn, one of the world's major grain crops and the most important crop in the United States, is an excellent example of the power of hybridization. Mass selection of the best samples of corn initially yielded a degree of positive improvement. However, in time no further improvements of nonhybrid or open pollinated corn came about. Corn is a monoecious plant and therefore is almost totally open pollinated. Consequently, corn is usually heterozygous for most genes. Scientists then turned to self-pollination, resulting in inbred lines of corn. Selected inbred lines were then crossed, and the resulting progeny produced the hybrid seed sold to the farmer. The resulting hybrid is characterized by hybrid vigor, often expressed in a plant that is more productive than its parents or open-pollinated varieties. When considering how hybrid seed is developed for corn, it becomes obvious why seed from a hybrid cannot be used to propagate another crop of corn.

Hybrid vigor occurs in many plants other than corn. It generally occurs in the heterozygous F1 generations following the crossing of plants that are homozygous or nearly so. Though the basis of hybrid vigor is not fully understood, several hypotheses may at least partially account for this phenomenon. Many crop plants in addition to corn have benefited from the process of hybridization. It is a particularly attractive approach to crop improvement.

The story of hybrid crops in the United States is not complete without mentioning the role of Vice President Henry Wallace, who began selling the first hybrid corn seed and was instrumental in starting the Hi-Bred Corn Company, then Pioneer Hi-Bred International, a subsidiary of the DuPont Company and now DuPont Pioneer. While the method leading to these developments is somewhat murky, Iowa State University and USDA Agricultural Research Service scientists were involved. It should be recognized that many scientists played a role in the hybridization of corn.[1] However, that Wallace was a visionary is not questioned. Wallace is probably the last president or vice president who was knowledgeable about and specifically interested in agriculture. It has been reported that while the USDA-ARS laboratory was being built north of College Park, Maryland, Vice President Wallace would, on occasion, take his lunch hour to drive out to see how the construction was progressing. It would be great to see a senior leader of the executive branch of our government take such an interest today. By 1940 hybrid seed had been planted on more than

90% of the corn crop in the Midwest. In Iowa, the planting of hybrid varieties approached 100%.

Hybrid corn was not planted in some regions of the United States until after World War II. Even then, it took a few years for many southern farmers to accept hybrid corn. The primary reason was that the available hybrids were not well suited to the southern environment. Those that were available were developed for use in the Midwest and were particularly susceptible to the rice weevil. Farmers also voiced their concern that since they could not save seed from the present crop for next year's planting, something must be wrong. In a few years, as varieties better adapted to conditions in the South became available and produced phenomenal yields, even the most reluctant farmers acquiesced.

Clearly, the concept of hybridization qualified as a major new innovation for agriculture. The really important aspect of hybridization is that production advantages of crops can be achieved with little or no additional costs. Even though not as far reaching as and lacking the magnitude of animal power, mechanical power, or commercial fertilizer, hybridization demonstrates that new innovations in agriculture are becoming more often based on science and the results of scientific investigation. This trend really began with commercial fertilizer. However, with classical genetics and hybridization, it became obvious that success was more often than not based on scientific studies. Obviously, the key to agriculture's future was an enhanced commitment to agricultural research.

Emergence of Science as a Basis for New Paradigms in Agriculture

Throughout much of the evolution of agriculture, developments of new paradigms were primarily practical solutions to various problems. The development of early hand tools was not based on a science-driven process. Food preservation was primarily a hit-or-miss system of development. For example, one can imagine that some of the days' kill could have inadvertently been left close to a fire overnight. After getting a good smoking, the hunter sampled the kill the next day and discovered a new way of processing meat rather than eating it raw. It is easy to visualize other such developments through serendipity. Perhaps some meat was inadvertently left outside, and conditions were just right for drying.

Preservation of food is one paradigm that continues to evolve to this day. Most industrial-era preserving processes are science based. The use of modified atmosphere or ionizing radiation as means of preserving food is clearly science driven. The early implementation of the concept of fertilization, such as the use of animal manures and bird guano, involved both innovation and serendipity. By treating phosphate sources with hydrochloric or sulfuric acid, this paradigm quickly evolved to a science-based process.

Early developments in mechanical power were primarily by innovation and basic engineering, whereas later ones are clearly based on scientific investigation. This is particularly true of guidance systems and various means of propulsion. Beginning with hybrid vigor and further genetic improvements of plants and animals, enhancements were heavily science based. Agrochemicals and all aspects of biotechnology are based exclusively on science. In considering the breadth of paradigms in agriculture, it is apparent that the trend is toward science-based improvements. It is a safe assumption that most future advances, certainly those leading to new paradigms, will be science based. This, of course, means that they will depend upon research.

5
Recent Paradigm Changes in Agriculture

*I am led to reflect how much more delightful to an
undebauched mind is the task of making improvements on
the earth, than all the vain glory which can be acquired
from ravaging it, by the most uninterrupted career of
conquests.*
—GEORGE WASHINGTON

Agriculture is an evolving process. From the previous chapter, it is
evident that future evolution is likely to be science driven. In this
chapter newer science-based paradigms are presented that reflect
the progression of agriculture and illustrate why greater support for
agricultural research is vital to our future well-being. Since around
the mid-twentieth century, developments in agriculture have been
almost totally based on basic science and technology. As this trend
continues, even more research will be needed. Further evidence of
this is the program established in 2009 by the National Science Foundation (NSF), which is searching for new challenges related to the
mission "to support innovative basic scientific research designed to
address key constraints to smallholder agriculture in the developing
world." This effort by the NSF is known as Basic Research to Enable
Agricultural Development (BREAD). Obviously, such an effort will
demand a greater investment in agricultural research.

Agrochemicals

Having observed the introduction of mechanical power and hybridization as a small boy, I became an actual participant in the agrochemical paradigm. Growing up on a farm, I could see the devasta-
tion caused by crop diseases, insect infestations, and losses caused

by weeds. In fact, I was very much a part of combating these difficult challenges to crop production and yields. Though there were few effective and available insecticides, we often resorted to actually picking worms off of some high-value crops such as tobacco. Weeds were another matter. They could be removed by hand but at very high costs. Hand removal of weeds from crops is one of the most backbreaking, onerous farm tasks. I recall, as a 10-year-old boy, hoeing (removing by hand) crabgrass in peanuts. If someone had told me that one day a chemical would get rid of the weeds (only one pint spread across an acre would kill all the crabgrass), I would not have believed it. To me that would have been much more far-fetched than a human going to the moon. These arduous farm tasks created an enormous appetite for and notable appreciation of the age of agrochemicals. My college education focused on agronomy with a specialization in weed science. Of course, herbicides and other agrochemicals were part of this field of study. In addition, my research career involved agrochemicals, primarily herbicides.

Chemicals have been employed as pesticides to protect crops dating back more than 4,000 years. Sulfur was probably the first chemical used as a pesticide. Others include arsenic, mercury, salt, lead, and naturally occurring materials such as pyrethrins and rotenone.

In the latter days of the nineteenth century, copper sulfate was found to be effective in controlling certain weeds in cereal crops. It was also observed that several fertilizer chemicals, including sodium nitrate, ammonium sulfate, and some potassium sulfates also had herbicidal properties. Bordeaux mixture, which contains copper sulfate and hydrated lime, has been used for more than 100 years to control fungal infestations in grapes. It is also effective in controlling certain diseases of potatoes, peaches, and apples.

While these and other examples illustrate the early use of chemicals in agriculture, it was not until the mid-twentieth century that we really saw the age of agrochemicals in agriculture blossom. One that really ushered in the agrochemical age was actually first synthesized in 1874. The insecticidal properties of this chemical, dichloro-diphenyl-trichloro-ethane (DDT), was discovered by Paul Müller, a Swiss chemist, for which he received the Nobel Prize in 1948 (*The Nobel Prize in Physiology or Medicine 1948*). It is difficult to overstate the impact of DDT on society at the time of its introduction. Not only did DDT play a key role in World War II, but it was also used extensively to control the insect vectors of typhus and malaria. At the end of the war this "wonder insecticide" was released for civilian use by farmers and homeowners. I vividly recall health authorities traveling throughout the South spraying homes with DDT. It was great sport to watch houseflies die before your eyes.

Of course, DDT was only the beginning. Many new areas of chemistry proved that other compounds also had insecticidal properties. Almost immediately after the release of DDT, concerns arose about the possible side effects of its use. However, because of the tremendous success of this compound as an insecticide, many of these fears were for the most part ignored during the early days. That is, until the publication of a book that shook the agricultural world: *Silent Spring*, by Rachel Carson, which was first published in 1962.

Silent Spring quickly became a best seller. In her book, Carson postulated that certain pesticides, including DDT, were harmful to the environment and endangered human health. After many months of wrangling and a study by the Presidential Science Advisory Committee, DDT and some other insecticides began a long phase-out. It is common knowledge that DDT was a monumental breakthrough that served humanity well (and still does in some situations in different parts of the world). However, it did so in two other important aspects. First, it showed the power of organic insecticides. Second, but equally important it uncovered the potentially deleterious side effects that can accompany some "wonder" pesticides.

The modern era of herbicides was ushered in with the synthesis of a simple chemical, 2, 4-dichlorophenoxyacetic acid (2,4-D). This chemical, when formulated as an amine, is highly selective. "Selective," as used here, denotes that some plants are more sensitive to the herbicides than others. The ester formulation, which is also widely used to control vegetation where an oil-soluble formulation is required, is less selective. The amine formulation kills many dicotyledonous (broadleaf) plants, but most monocotyledonous (grass) plants are not affected. Developed during World War II, it was released for commercial use in 1946. It proved to be a highly selective herbicide and was widely used to control broadleaf weeds in grass crops such as corn, wheat, rice, and other cereals. But 2, 4-D was only the first phenoxy acid herbicide. Soon others with different properties and a different range of selectivity (e.g., 2, 4DB and 2, 4, 5T) were synthesized.

Just as DDT showed what was possible with organic insecticides, 2, 4-D was an excellent example of what the herbicide age could bring about. Most herbicides as well as other agrochemicals were synthesized in industrial laboratories. This effort exploded in the 1950s, '60s and '70s. At one time there were more than 50 industrial laboratories in the United States and other countries actively synthesizing and looking for chemicals in the search for fungicides, insecticides, herbicides, and other chemicals with unique pesticidal properties. Today, there are fewer than a dozen. The benzoic acids, chloroacetic acids, substituted ureas, s-triazines, bypyridyliums, ammonium salts, carbamates, and many others soon followed the introduction of the phenoxy herbicides. Having been a weed science researcher and teacher during the unfolding of these new developments, I can say the success brought a sense that one of humankind's oldest plagues (weeds) was almost "out for the count." Unfortunately, we overlooked the power of biological organisms to adapt!

The third of the "big three," fungicides, evolved in a fashion somewhat similar to that of insecticides and herbicides; however, there was no "eureka" moment with fungicides. The problem with plant diseases probably began along with agriculture and is still with us. Early on, the causes or causative agents of plant diseases were unknown or else attributed to the environment or some unknown factor. It was not until the early nineteenth century that the idea of a causal agent was accepted as accounting for plant diseases. Plant diseases have played a central role in human history. Some of the smut and rust diseases of the grains, particularly corn, have had a significant impact on people. Of course, late blight of potato caused a major upheaval in Ireland, which resulted in a mass migration, particularly to the United States.

Weeds and some insect problems could be dealt with by employing known technology such as the plow and the hoe, both involving hand labor. However, dealing with plant diseases was essentially impossible. Consequently, the time was ripe for the introduction of chemical control of plant disease. Fungicides that control fungal diseases are only one group of the pesticides used in the battle against plant pests. We also should recognize nematicides for control of nematodes, algicides for control of algae, and miticides for control of mites.

Introduction of agrochemicals provided yet another major new paradigm for agriculture. Their impact on agricultural productivity was immense, and many of the agrochemicals introduced in the latter half of the twentieth century made life better for people. Herbicides reduced or often eliminated one of the more unpleasant aspects of agriculture: hand labor. Insecticides enhanced the quality of life by controlling the vectors of many diseases.

While the pace of synthesizing new pesticides has slowed in the past two or three decades, interest remains in identifying new and more effective chemicals. One such pesticide, the chemical glyphosate (N-phosphonomethyl glycine), was recognized as having herbicidal properties in 1970. Because it is phytotoxic to many green plants, glyphosate controls a broad spectrum of weeds, making it a highly useful herbicide. However, scientists at the Monsanto Company, who initially recognized the efficacy of this herbicide, reasoned that if certain crop plants could be genetically engineered to be resistant to the herbicide, its use would be greatly enhanced and expanded. That is precisely what Monsanto scientists did. Today, glyphosate, whose original US patent expired in 2000, is still the most widely used herbicide in the United States and in much of the world. Engineering desirable crop plants to be resistant to a herbicide is truly a revolutionary approach to dealing with one of the major challenges in agricultural production. It is almost a new paradigm in and of itself. It has fostered a marriage of seed technology, genetic engineering, and agrochemistry, thus fostering mergers of major seed and agrochemical industrial companies.

Biotechnology

Biotechnology is but one step in the progression of agricultural science. Each of the steps brought about a new agricultural paradigm that enhanced the production of food, which, in turn, freed up a significant portion of the population for other pursuits. In the first half of the twentieth century additional innovations included hybrid corn, the introduction of an array of agrochemicals, mechanical power, and the era of biotechnology. Bioenergy is a more recent development in agriculture and has the potential to compete with food production. Effectiveness of agricultural research will greatly influence the outcome.

The era of biotechnology was made possible by a number of significant developments in the basic sciences. In fact, one of the most important discoveries in the biological sciences in the twentieth century—or perhaps of all time—was the discovery and description of the double helix structure of deoxyribonucleic acid (DNA). The article announcing the finding, "A Structure for Deoxyribose Nucleic Acid," was authored by James D. Watson (an American) and Francis Crick (an Englishman) and was published in the journal *Nature* in 1953 (Watson and Crick, "Structure for Deoxyribose Nucleic Acid").

The elucidation of the structure of DNA and, in particular, how it can replicate itself had a profound impact on the biological sciences in general and on agriculture in particular. As in so many discoveries in science, major discoveries are the result of the efforts of numerous unsung and unrecognized contributors. The description of the structure of DNA certainly was due to the work of a number of background researchers. This new understanding of the structure of DNA opened many avenues to further investigation. First, it revealed that the genetic description of an organism could be replicated and passed on to a succeeding generation. Such information is encoded in a gene, a fragment of DNA, and is the basic unit of heredity in living organisms.

Though much of this investigation was done outside the traditional agricultural research community, agricultural scientists were quick to recognize the potential of biotechnology for agriculture. In their book *Agricultural Biotechnology*, published in 1987, Hess et al. explain their visualization of the potential power of biotechnology. They cite techniques such as "recombinant DNA, gene transfer, embryo manipulation and transfer, plant regeneration, cell culture, monoclonal antibodies, and bioprocess engineering" (ibid., p. 3).

Hess et al. further state that "biotechnology offers tremendous potential for improving crop production, animal agriculture and bioprocessing. It can provide scientists with new approaches to develop higher yielding and more nutritious crop varieties, improve resistance to diseases and adverse conditions or reduce the need for fertilizers and other expensive agricultural chemicals. In animal agriculture, its greatest immediate potential lies in therapeutics and vaccines for disease control" (ibid.). During the years since publication, developments have proven the authors were on target with their expectations of biotechnology.

I was serving as a director of an agricultural experiment station during these early days of the application of biotechnology in agriculture. I vividly recall that one day a group of experiment station scientists submitted a request for a significant sum of experiment station money to start some research in this area. Money was made available, and after a few months I received news that they were making good progress. I requested more details and made a visit to their laboratory, where I found out that they were working on two plants, Arabidopsis and petunias—hardly major crops for my state or any state. But such was common, particularly in the early days of biotechnological research.

Even in the very early days of these developments, many agricultural scientists perceived this area of science to be one of the most important new paradigms in agriculture's history. The first page of the National Academy of Sciences book *Agricultural Biotechnology* recognizes this: "Agriculture has moved from a resource-based to a science-based industry as science and technology have been substituted for land and labor" (Hess et al., p. 1).

Among the first fields to capture the power of biotechnology was medicine. The amount of glucose in the blood of humans is regulated by the hormone insulin, which is produced in the pancreas. Insulin is released in response to blood glucose. When this process is not working properly, a person has diabetes. A person who suffers from diabetes has serious health problems, including fatigue, constant infections, blurred eyesight, numbness, tingling in the hands or legs, increased thirst, slowed healing of bruises and cuts, and circulatory problems. Of course, if a person suffers from diabetes, the solution may be to take insulin. After identifying and purifying insulin from dogs, scientists discovered that the structure of insulin isolated from the pancreas of cattle and pigs was quite similar to that of human insulin. For several decades cattle and pigs served as sources of insulin for diabetics. Then, in the 1980s, scientists were successful in employing recombinant DNA technology to identify the genetic information necessary to produce insulin. This information was put into bacteria that could make human insulin. Today, many diabetics throughout the world have benefited from this process. Insulin from genetically modified organisms (GMOs) was almost immediately accepted by diabetics.

In 1994 the power of biotechnology in agriculture really began to be employed with genetically modified food, such as the "Flavr-Savr" tomato in the United States. Since this rather inauspicious beginning, rapid growth and the adoption of genetically engineered crops have ensued in the United States (see figure 2) (Fernandez-Cornejo, "Rapid Growth"; Fernandez-Cornejo and Caswell, *First Decade of Genetically Engineered Crops*). Genetically modified foods have been developed to take advantage of effective resistance to insects and adverse growing conditions and to exhibit herbicide tolerance. Genetic modification can lead to improved nutrition, new pharmaceuticals, and even pollution abatement.

Resistance to Insects

The insertion of a gene from *Bacillus thuringiensis* (Bt) enables cotton and corn plants to produce a highly toxic, but specific, targeted protein that kills only insect pests. This saves numerous applications of insecticides during a typical growing season and reduces the carbon footprint of crop production as well. The Bt gene has been inserted in a number of crops with similar success. This concept has great potential for solving a number of serious insect as well as disease problems in crop species around the world (Fernandez-Cornejo, "Rapid Growth"). Even though strategies were employed from the beginning to mitigate insect resistance, the development of insect resistance is a constant threat and must be carefully managed to ensure a continued successful approach.

Herbicide Tolerance

The Monsanto Company released herbicide-tolerant (HT) crops in the 1990s. The most widely used HT crop, soybeans, suffered no ill effects or yield losses when sprayed with the highly effective herbicide glyphosate (Roundup). In addition to soybeans, the Roundup Ready gene has

Figure 2. Adoption of genetically engineered crops in the United States, 1996– 2013.*

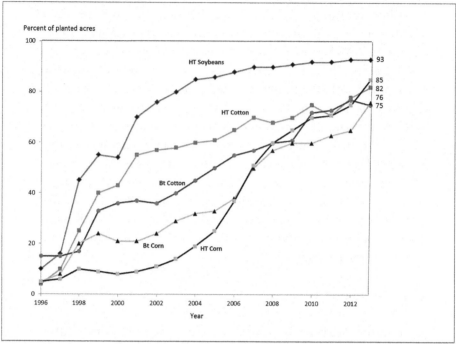

*Fernandez-Cornejo, "Adoption of Genetically Engineered Crops in the U.S."

been inserted in a wide range of major crops, including cotton, corn, and canola. This concept has been employed to develop other herbicide-tolerant crops. While developing resistance is an ongoing concern, different research approaches are seeking to minimize this challenge by developing new technologies for the elimination of transgenes from pollen and seeds. The approach of using genetically modified crop plants that tolerate herbicides has great potential to address the control of certain weed species in crops. Of course, the challenge of herbicide resistance developing in some weed species remains.

Resistance to Adverse Growing Conditions

Another area that holds great promise for GMOs is the development of plant cultivars with unique attributes that enable plants to thrive under adverse growing conditions. This is especially important in view of the potential for global climate change. Regardless of the nature of the climate change, there is a need for plants that thrive in hot, dry, and cold climates, with elevated levels of CO_2 and in conditions of excess as well as diminished levels of moisture, at various levels of salts, and so on.

Predicted consequences of global climate change include flooding and drought. Since 1971, various countries and organizations have been supporting basic and applied research on the genetic improvement of rice at the International Rice Research Institute (IRRI) through its membership in the Consultative Group on International Agricultural Research (CGIAR). Collaboration between IRRI and the University of California led to the identification of a gene in certain cultivars of the genus *Oryza*, which enables rice to tolerate complete flooding for up to two weeks (Xu et al., "*Sub 1 A*." This trait has been successfully incorporated into currently grown cultivars of rice.

In Japan, scientists have pioneered basic research on drought tolerance in plants and are starting to see major payoffs from their efforts. It is anticipated that in a few years crop cultivars that have some degree of drought tolerance will be developed and released. This will have great implications for much of the world. Not far behind are engineered cultivars that are more efficient users of plant nutrients.

Nutritive Benefits

The development of "golden rice," which has been modified to have a high provitamin A content, is an excellent example of the potential for use of GMOs to make plants more nutritious. As we learn more about human nutrition, we are finding countless opportunities for designing foods specifically for nutritional requirements.

While much of the early research was conducted with plant models, there was great interest in animal species. Considerable research was conducted by the Alabama Agricultural Experiment Stations in developing transgenic fish in an effort to improve productivity. Since this early investigation, numerous studies have employed the tools of biotechnology to improve animal species.

These few examples illustrate the potential biotechnology holds for agriculture. Indeed, it is inconceivable that we can meet the future demand for food, feed, fiber, and fuel without the tools of biotechnology. Introduction of genetically modified food has not been without controversy. Though almost no adverse response to insulin produced by genetically modified bacteria has occurred, genetically modified food was widely rejected in the early days of biotechnology. At the present time food derived from genetically modified crops permeates the entire food chain in the United States. Much of the corn, cotton, and soybean crops are grown with genetically modified seed. Thus, in reality there is an ongoing "experiment" with more than 300 million test subjects in the United States. So far there have been no reported adverse observations.

It is noteworthy that each of the major innovations in agriculture in the twentieth and twenty-first centuries has been highly controversial. I rather suspect that some of the earlier innovations were also disputed. One can only imagine the reluctance of some when a water buffalo was first yoked to pull some type of crude implement. As previously discussed, hybrid corn was not readily accepted by a number of farmers. The introduction of agrochemicals has been highly divisive and is still being debated. And while we are in the very early stages of energy production by using biomass to produce biofuels, the uproar about food versus fuel is a reflection of the sensitivity of this approach.

Use of Biotechnology to Improve Animal Agriculture

While much of the early research effort employing the tools of biotechnology to improve agriculture focused on crops, the use of biotechnological techniques to develop transgenic animals has also been recognized as having great potential. A transgenic animal is one that has undergone a specific modification of its genome. This is done to provide some desirable trait such as faster growth or to produce more milk in cattle or wool in sheep. It may also be possible to develop transgenic animals that possess greater resistance to certain diseases. One of the most exciting opportunities is the development of transgenic animals that can be used as disease models or to produce some specific product such as useful human medicine.

Animal cloning and genetic engineering technology along with use of embryonic stem cells have great potential for addressing some of these challenges in animal agriculture. Developing means of enabling food animals to be resistant to many animal diseases such as Newcastle virus in chickens would usher in a new paradigm in agriculture. These approaches may also help develop treatment for neurodegenerative (e.g., Parkinson's and ALS) and vascular diseases. Clearly, most innovations that brought about substantial improvements in agricultural productivity were disputed. Understandably, the use of biotechnology and subsequent developments have been highly controversial, and this debate is still ongoing in some parts of the world.

Agriculture as a Producer of Energy

Each of the new paradigms described in this chapter brought about major changes in agriculture. It is now becoming apparent that agriculture will be involved in meeting the challenge of securing a sustainable energy future. Indeed, for the people of this planet to survive, each and every person must have adequate food and energy. In the past, we thought of agriculture as our source of food for humans, feed for our livestock and companion animals, fiber for our clothes and home, and flowers for our landscape and environment. In the early years of the twenty-first century, all but the most foolish realized that the future of our civilization depends on our ability to achieve sustainable food and energy security for everyone; hence the convergence of agriculture and energy.

Food and energy are absolutely necessary for the existence of our civilization as we know it. There are "experts" on every side of the food and energy issue. Martenson, in *The Crash Course*, paints a rather bleak picture of our energy future. Conversely, Laughlin, in *Powering the Future*, is optimistic about energy security. Since no one knows precisely what the future will bring, in this situation common sense is far more important than the opinions of experts. The single most important fact is that fossil energy is a finite resource. Fossil energy, including natural gas, coal, and petroleum, comes from plant material accumulated over a period of millions of years. Basically, they are energy accumulations derived from the energy of the sun. Another inescapable fact is that each of these energy sources is finite.

This means that they will become increasingly scarce and one day will be exhausted.

Sheikh Zaki Yamani of Saudi Arabia captured the essence of the forthcoming transition we must make beyond the Oil Age when he said, "The Stone Age did not end for lack of stone, and the Oil Age will end long before the world runs out of oil" ("The End of the Oil Age"). The Stone Age ended because better alternatives to stone were discovered—namely iron, steel, gunpowder, and so on. We must accelerate our efforts today to find other and more successful alternatives to oil and other sources of fossil energy. Considerable effort has been directed at capturing wind energy, geothermal energy, hydropower, nuclear energy, and the energy in ocean and river currents. Each of these has some potential and can play an important role. However, the sun probably holds more promise than any other potential energy source. This assumption puts agriculture squarely in the middle of achieving sustainable energy security. Capturing the sun's energy by plants is the fundamental basis of agriculture.

The sun is the center of our solar system and, as described in chapter 1, holds great promise for addressing the future energy needs of the planet. The fossil energy sources (e.g., coal, petroleum, natural gas, tar sand) that are so useful today are accumulations of the sun's energy collected over millions of years. Our challenge today is how to capture the sun's energy in a much shorter time frame. Considerable research efforts have been expended on developing means of improving our ability to capture the sun's energy. Solar energy can be converted into electricity, heat, and biomass. Photovoltaic cells (solar cells or panels) can convert the sun's energy directly into electricity. Another important approach is employing solar energy to heat a liquid that in turn creates steam, which turns a traditional turbine to generate electricity. The sun's energy can be used to provide thermal energy, which can be used as a heat source. This was a very effective means of heating water in the western United States as early as the 1800s.

Solar lighting is another approach to capturing the sun's energy. Solar lighting is finding widespread use in yard and pool lights. Considerable effort has been devoted to developing means of using solar tubes to bring sunlight into buildings.

Another important approach is to capture the sun's energy in plants. Biomass production is possible because of an extremely important chemical process—photosynthesis. Photosynthesis is an elegant biochemical process occurring in green plants where, in the presence of sunlight, carbon dioxide, and water, yields organic compounds such as glucose and oxygen and a host of secondary products. This process is the ultimate source of all food, fiber, and much energy on Earth. It is the foundation of agriculture.

Why does biomass hold such great potential as a renewable energy resource? What are the advantages for the production of biomass through photosynthesis, and how do they compare with other renewable energy sources such as wind power, hydropower, geothermal energy, and the energy in river and ocean currents and ocean waves? Each of these renewable energy sources primarily is a producer of electricity. Also, like fossil energy, the availability of wind power, hydropower, and geothermal energy is not universal. On the other hand, biomass can be produced in most regions of the world and converted into liquid fuels, which are very important for transportation. Biomass is important because of the immense amount of energy trapped by photosynthesis.

Second, photosynthetic energy captured by green plants and stored in seeds and other plant parts may have many uses. Biomass can be the feedstock for liquid fuels or thermal energy or be converted into bioproducts. Third, the process of biomass production is based on a tried and true technology of the sun-driven chemical process—photosynthesis. It is one of the simplest chemical reactions in plant science and is of the utmost importance to humankind (figure 3).

The formula for photosynthesis in green plants involves carbon dioxide in the air and water in the soil, both in the presence of sunlight, and leads to the production of oxygen and glucose. Capturing the energy of the sun through the production of biomass holds great promise for our present and future energy needs. But it is not all smooth sailing. An effective and cost-efficient means of converting cellulosic material to some form of usable energy source, such as a transportation fuel, must be available. While there is considerable ongoing research effort, much misinformation is clouding the potential of this process. I am

Figure 3. Chemical equation for photosynthesis.

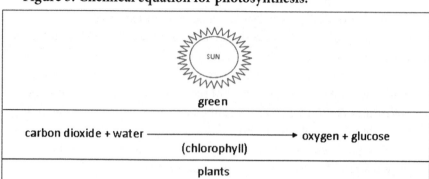

aware of several proposed processes. Even though many approaches are currently being researched, no one has yet made a cellulosic transportation fuel that is competitive with a petroleum-based fuel. Indeed, this remains a major research challenge.

Sustainable energy security must be achieved at some point in the future. Consider Brazil's accomplishments, which provide some encouragement. Brazil has achieved energy security by basing much of its transportation fuel needs on ethanol derived from sugarcane. However, the United States is the world's largest producer of ethanol, which is derived primarily from corn. The progress made to date employs both technology and crops that go back thousands of years. These crops (corn and sugarcane), using yeast fermentation technology, which dates to the early Sumerians and Egyptians, have accounted for most of the success to date in sustainable energy. Some magic solution to the planet's energy challenge may yet exist; however, in the absence of a miracle, we must plan and build on technology we have, can develop, or at least can visualize.

As we consider our survival on this fragile globe, it is apparent that we must have adequate food, fiber, and energy. Agriculture is the primary source of food and fiber and, in all probability, will be one of the significant sources of energy. Responding to these needs and expectations is one of the most important challenges of our time. My greatest concern is that we as a nation and the entire world are not taking the most important approach to addressing the problem.

We are not investing in research commensurate with the challenge of either food or energy production.

Recognizing that generating energy is an integral part of the agricultural mission brings up yet another major challenge for agriculture. According to the second law of thermodynamics, energy cannot be created. Energy can only be changed in form—unfortunately, always at a cost (Hess et al., *Agricultural Biotechnology*). Addressing this new paradigm will require an unprecedented commitment to research.

The sun is our only hope for producing our food and fiber and is among our best approaches to supplying much of the energy we need to power our civilization. Of the ways of capturing the sun's energy, green plant photosynthesis is the most promising. This puts agriculture at the very heart of our future survival on this planet. Agriculture is, indeed, the key to our survival. The success of agriculture, in the end, will be determined by information, knowledge, and technology gained through research. Clearly, agricultural research must be considered as key to our future and thus deserves a fair share of the federal research dollar (Mervis, "Battle over the 2011 Budget").

Each of the major developments discussed in this chapter has brought about significant changes in agriculture. The selection of each was somewhat arbitrary, and one could make a case for other developments. Obviously, many other innovations in the long history of the evolution of agriculture could also be included. It is obvious these innovations were not clear-cut, sharply defined changes. Some unfolded over a period of decades, while others are still unfolding, with great expectations yet to be fulfilled.

From human beings' early beginning as hunters and gatherers to the recognition of the importance of grains in making bread and beer (as described by Sinclair and Sinclair in *Bread, Beer and the Seeds of Change*), agriculture has steadily evolved. Many agricultural innovations that brought about new paradigms occurred during the twentieth century. As mentioned earlier, I personally experienced the transition from animal power (mules) to tractors and hybrid seed corn on my family's farm. My teaching, research, and administrative careers embraced the age of agrochemicals and biotechnological developments. Moreover, I was working in the US government when it became clear that agriculture would play a role in meeting the energy challenge. This latter is perhaps the greatest challenge that agricultural research has ever encountered.

It is interesting to speculate about what the next paradigm change in agriculture will be. It will be science based and the result of innovative and highly creative research. It will definitely not come about by timid incremental research. The evolution of agriculture as described in chapters 4 and 5 clearly points to the future of agriculture and will be achieved only through dedicated, innovative agricultural research.

Meeting the Challenges Associated with Science-Based Agriculture

As pointed out in the previous chapter, new agricultural paradigms will most likely be based on science. Forewarned should lead to being forearmed. This is not a minor issue in a society that is already somewhat science challenged. Many additional challenges are exceedingly complex and thus beyond the knowledge base of a layperson. In today's society a number of groups have a particular interest in and take positions on issues that are not consistent with truth and relevance. The power of the media can be extraordinarily helpful as an educational tool. On the other hand, it can sow seeds of doubt. The following quote appeared in an article in a popular magazine: "Fertilizer should never have been allowed in agriculture. I think it's time to ban it. It's a weapon of mass destruction. Its use is like war, because it came from war" (Vandana Shiva, quoted by Specter, "Seeds of Doubt"). One does not have to be gifted intellectually to identify the errors in this statement. But one is left with the idea that the use of fertilizer is not only unnecessary but could also be harmful. Of course, misuse and overuse could cause problems, but clearly current yields could never have been achieved without fertilizer. I cannot imagine anyone who has any training in or association with agriculture agreeing with the statement. It is absolutely absurd.

Another interesting case was the cranberry crisis of 1959, which was precipitated by amitrole, a nonselective triazole herbicide. Normal use in accordance with the label would leave no residue; however, improper use could result in a measurable residue. Thirteen days before Thanksgiving in 1959 the secretary of health, education, and welfare announced that a portion of the current crop of cranberries showed residues of amitrole. Even though the levels were very low, a near panic ensued. Housewives were throwing out cranberry sauce made before amitrole was even on the market. Since amitrole did have carcinogenic effects, its use was prohibited by the Delaney clause (a provision of the Food Amendment Act of 1958), which forbade the presence of any amount of carcinoid substance in food. Though it was difficult to put the genie back in the bottle, after further study and deliberation, a plan was implemented to certify some cranberries to meet holiday needs.

As mentioned earlier, one of the most challenging science-based issues affecting agriculture in recent years is that of GMOs. Today, not only the United States but also other developed and developing countries are rapidly accepting GMOs. With regard to their use, considerable concern has arisen about the safety of both humans and the environment. Perhaps it is because we have all lived through the introduction and development of GMOs that we appreciate the concern. I am not sure such products could have been introduced without controversy. As already pointed out, most innovative technologies in agriculture have created anxiety among some people. Throughout history people have had to weigh the risks and the benefits of any technology. While GMOs have made considerable contributions already, many great opportunities lie in the future with the judicious investment of research funds.

Since current agriculture is heavily science based, hardly any aspect escapes close scrutiny. Over the past five decades a number of science-based challenges have had sufficient merit to warrant careful evaluation. Some of these include antibiotics in animal feeds and Alar (trade name for daminozide) use in apples in 1989. Issues such as those mentioned here are heavily science based; thus addressing them requires a dedicated commitment to the basic principles of science.

To reach a proper resolution of such issues, all parties must accept responsibility. The agricultural research community must address these matters in a rigorous manner without regard to political correctness. On the other hand, the questioning public should strive to appreciate the scientific process. The media can play a constructive role by engaging in honest, fair, and balanced reporting on such topics. This can and should be done without sensationalizing the issue.

In view of the fact that most of the new paradigms affecting agriculture in the future will be science based, research will be very important in meeting the challenges of tomorrow. *Our future survival on this planet is contingent on the appreciation of science and the success of the agricultural research effort.*

6
Success of Agricultural Research

Being busy does not always mean real work. The object of all work is production or accomplishment and to either of these ends there must be forethought, system, planning, intelligence, and honest purpose, as well as perspiration. Seeming to do is not doing.

—THOMAS A. EDISON

There are multiple ways to account for the success of agriculture. Probably the best measure is the population that can be supported by the output of agriculture. When formal agricultural research was just beginning, the population of the planet was about 1.3 billion. One hundred years later—in 1950—it had doubled to about 2.5 billion. Only 50 years later—in 2000—it had increased to 6 billion. In only the last 14 years the world's population has grown another 1.3 billion. It is not fortuitous that population growth is so closely related to the emergence of agricultural research. This close association between them and agricultural output is to be expected.

Only 170 years have elapsed since the founding of the Rothamsted Experimental Station in England, which marked the beginning of formal agricultural research. In this brief period, the success of agriculture made possible largely by agricultural research has been nothing short of miraculous. Admittedly, that success is a reflection of many factors. However, research establishes the potential for success. Without the necessary information, knowledge, and technology, other contributors would be marginalized.

When speaking of accomplishments in agricultural research, the immediate question is, just what constitutes a measure of success? The standard definition of an accomplishment is something done successfully—an achievement. This definition applies to and exemplifies agricultural research. Accomplishment also contributes to a desirable outcome for or an expectation of humankind. Of course, it is left to the nature of the research to define the value of an accomplishment and its contribution to success. Most accomplishments in agricultural research have multiple values. For example, research to develop a mechanical device for harvesting a vegetable crop would be greatly beneficial to production and may also eliminate the backbreaking drudgery of stoop labor.

There are numerous means of assessing the value of specific accomplishments in agricultural research, many of which fit into one of the following categories:

- Increasing yield of agricultural output
- Elimination or reduction of drudgery experienced by those who work in agriculture
- Improvement in quality of output
- Improvement in quality of life
- Reducing the cost per unit of output
- Intangible contributions

Requirements for Successful Agriculture

Describing the accomplishments of research that contribute to agriculture's success is a bit like giving a history of the world. That the agricultural enterprise has been phenomenally successful is hardly a debatable issue. What is more problematic is what accounts for success. Observing the success of agriculture in various countries illustrates some of the many factors that contribute to agricultural achievements.

The first requirement involves the availability and quality of land. Though some deficiencies in land quality can be overcome with proper amendments and plant nutrients, these are added costs. Proper drainage is necessary for some land. Available water, particularly inexpensive and high-quality water, is another necessity. Access to modern farm equipment adds to the probability of success. In addition, a favorable climate and acceptable topography and geography are pluses. Availability of accessible markets also contributes to the success of agriculture. While sometimes overlooked, a favorable regulatory climate, good transportation system, and access to capital are often requirements. Of course, access to information, knowledge, and technology ensures agricultural success. And finally, good farmers who have the skills to manage the total portfolio can make the system work.

Much of early classical research dealt with investigations that did not lend themselves to easy validation. Certainly, Galileo's identification of planets was great research that produced wonderful findings, but they could not be validated at the time. Research in many fields began as a quest to understand a particular phenomenon such as the movement of the sun, moon, and stars and how the universe functions. On the other hand, agricultural research was initiated to solve immediate and specific problems. Solving the problem validated the research effort.

Legislation that created the state agricultural experiment station system in the United States provided for each such facility to develop its own research thrusts. Consequently, studies to address specific problems (applied) were a priority from the very beginning. Many individuals recognized that these institutions could address many aspects of agriculture requiring both basic and applied research.

Agricultural research has contributed in ways other than solving specific agricultural issues. Much of the early research on statistical methods was done in agriculture. Agricultural research was an early adopter of statistics for evaluating treatment effects. Because many treatments had effects that were not readily discernible, the tools of statistics enabled the investigator to distinguish such effects with a certain degree of confidence. This chapter presents a general description of some of the accomplishments that validate the success of the agricultural research effort, particularly in the United States.

Increasing Yield of Agricultural Output

An important means of evaluating accomplishments in agricultural research is to consider enhanced productivity (i.e., increase in yield, output, or growth). Obviously, at the outset research cannot account for all of the positive impacts of productivity. However, agricultural research is responsible for establishing the potential for yield output. Factors such as improvements in machinery associated with enhancing yields have their beginning in a research process. Selected examples of yield improvements of commodities illustrate the effectiveness of research for enhancing productivity.

Corn

If the United States had a national crop, it would likely be corn, as it is grown in every state in the Union. The National Agricultural Statistic Service (NASS) has yield records for corn from 1866 to the present. This dates to almost the beginning of formal agricultural research.

Yield (national average) of corn in the United States in 1866 was 24.3 bu/A with a total production of 730,814,000 bushels. For the next 75 years, considerable variability in yields was recorded. After 1941, corn yields were consistently above 30 bu/A and climbed rapidly into the 40- and 50-bu/A range during the next couple of decades. Increases in per acre yield followed the introduction of hybrid corn in the '30s and widespread planting of these new varieties in the 1940s (figure 4).

The next breakthrough was achieving the 100-bu/A level, which occurred in 1978. Since 1978 only two years, 1980 and 1983, have produced yields of fewer than 100 bu/A.

Production of corn in the United States rose to more than a billion bushels in 1870. A 2-billion-bushel crop was recorded in 1885 and for most of the next 75 years. Total production increased rapidly, reaching 3 billion bushels in 1956, 4 billion in 1963, 5 billion in 1971, 6 billion in 1976, and 7 billion bushels only 2 years later. We had an 8-billion-bushel crop in 1981 and 1982 and reached 10 billion bushels in 1994. In 2004 the United States produced 11.8 billion bushels and 12–13 billion bushels each year through 2011. The nation accounts for 35–45% of the world's production of corn. This is somewhat lower than in recent years primarily because of increasing output in some countries, particularly China, Brazil, and Ukraine. Also, the United States has experienced drought in several major corn-producing areas in recent years.

Figure 4. Yield of corn by year, 1870–2010 (data plotted at 10-year intervals).

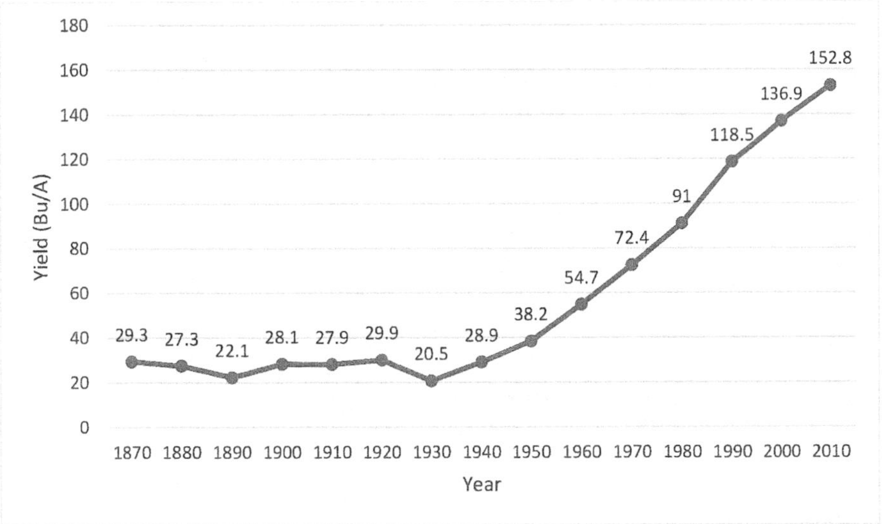

USDA National Agricultural Statistics Service historic databases. See http://www
.nass.usda.gov/Quick_Stats/.

Soybeans

The second most visible and important agronomic crop in the United States is soybeans. Indigenous to China, soybeans were introduced into the United States in the early twentieth century. Growers soon learned that soybeans were well adapted to the soils and climate of the Midwest and mid-South as well as many other crop-producing regions of the country (figure 5).

In 1924 NASS began keeping records on soybeans. In that year, yield was 11 bu/A with a total production of 4,947,000 bushels. Soybean yields were consistently above 18 bu/A beginning in 1945, with the exception of 1947, when the yield was 16.3 bu/A. In the mid-1980s, yields were generally in the 30-bu/A range and were in the 40-bu/A range in 2004.

Total production was a modest 10,000,000 bushels or less until 1930. Production grew rapidly in the next decade, reaching more

Figure 5. Yield of soybeans by year, 1930–2010 (data plotted at 10-year intervals).

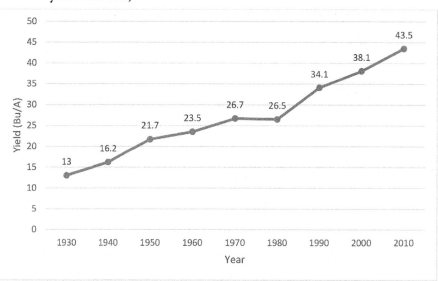

USDA National Agricultural Statistics Service historic databases. See http://www .nass.usda.gov/Quick_Stats/.

than 100 million bushels in 1941. Total production quickly grew to 200 million bushels in 1946; 300 million bushels in1954; 449 million bushels the next year; and more than 500 million bushels in 1958. Soybean production was 1.1 billion bushels in 1968 and remained above 3 billion bushels after 2008.

The United States produces more than half of the world's most important oilseed crop—soybeans. Almost 80% of edible oils consumed in the United States are derived from them. Soybeans find their way into numerous other food products and into nonfood uses such as protein for livestock, anticorrosion agents, soy diesel, and even waterproof cement.

Cotton

Cotton is grown in the southern tier of states from California to Georgia. The earliest year of records (1866) compiled by the NASS showed a yield of 122 lb/A. Yields remained in the 100–280-lb/A range until 1953. For the next three decades cotton yields fluctuated between 300

Figure 6. Yield of cotton by year, 1870–2010 (data plotted at 10-year intervals).

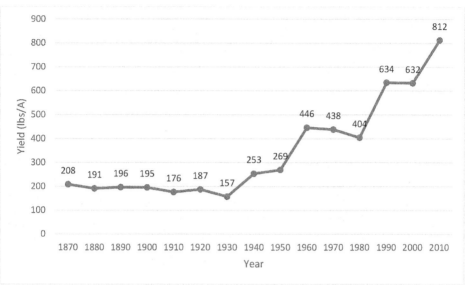

USDA National Agricultural Statistics Service historic databases. See http://www
.nass.usda.gov/Quick_Stats/.

and 600 lb/A. From 1984 to 2004 yields ranged in the 600–800-lb/A
range. With the exception of 2009 and 2011, yields were above 800
lb/A (figure 6). Production of cotton grew from slightly more than 2
million bales in 1866 to a high of 23,890,200 bales in 2005. Thereafter,
production declined and was 17,009,900 bales in 2012.

In 1866 7,660,000 acres of cotton were harvested. The high point
for harvested acres of cotton came in 1925 with 44,356,000 acres.
However, in 2011, only 9,849,500 acres were harvested. It is remarkable that total production was about 16 million bales in both years,
a phenomenon that can be explained primarily because of cotton research and effective implementation of research findings.

Peanuts

Another crop that is well adapted to the United States, particularly in
the Southeast, is peanuts. Yields of peanuts have increased from 660
lb/A in 1909 to more than 1,000 lb/A in 1953; 2,000 lb/A in 1970; and

Figure 7. Yield of peanuts by year, 1910–2010 (data plotted at 10-year intervals).

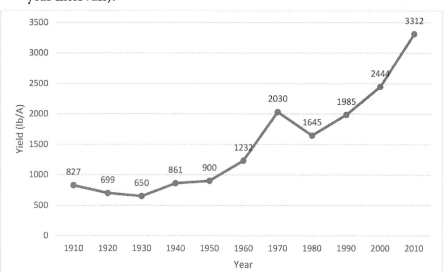

USDA National Agricultural Statistics Service historic databases. See http://www.nass.usda.gov/Quick_Stats/.

3,000 lb/A in 2010 (figure 7). Average yield was more than 2 tons/acre in 2012.

In the United States, total production of peanuts grew steadily from 354,605,000 pounds in 1909 to more than 4.16 billion pounds in 2010. Although production dropped to below 4 billion pounds in 2011, it reached more than 6.7 billion pounds in 2012. The success of this commodity can be accounted for by several factors; for example, genetics and breeding research led to new varieties of peanuts with greater levels of pest resistance to important insects and diseases and better and more efficient harvesting techniques. Also contributing to this achievement were more effective means of assessing maturity, which is complicated because the harvested crop is under ground, enabling peanuts to be harvested at the optimum time.

Figure 8. Yield of wheat by year, 1870–2010 (data plotted at 10-year intervals).

USDA National Agricultural Statistics Service historic databases. See http://www
.nass.usda.gov/Quick_Stats/.

Wheat

Commodities such as wheat have shown dramatic yield increases. Yield
of all wheat increased from 11.0 to 46.3 bu/A between 1866 and 2010
(figure 8). During this same period, total annual production increased
from 169,703 million bushels to 2.269 billion bushels. Even though the
United States produces only 9% of the world's wheat, it supplies about
20% of the world's wheat export market.

Barley

Yearly yields of all types of barley varied widely but stayed in the range
of 20–30 bu/A until 1960. Thereafter, yields quickly grew to more than
40 bu/A and were more than 50 bu/A in 1979 (figure 9). Yields stayed
above 50 bu/A and reached 60 bu/A in 1998 and 73 bu/A in 2009.

Figure 9. Yield of barley by year, 1870–2010 (data plotted at 10-year intervals).

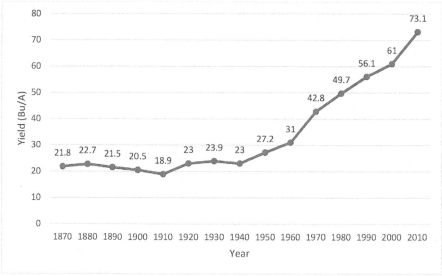

USDA National Agricultural Statistics Service historic databases. See http://www
.nass.usda.gov/Quick_Stats.

Tobacco

Yield of flue-cured tobacco (Type 14) (Georgia-Florida Belt) was less
than 1,000 lb/A until 1938. After 1946 yields were consistently above
1,000 lb/A and reached more than a ton per acre in 1965 (figure 10).
With the exception of 1976, 1979, and 2005 yields were above 1 ton/A
after 1974.

Young Chickens

Increases in average live weight and in total live weight of birds produced illustrate improvements in an animal enterprise. Average live
weight of young chickens was 3.35 lb/bird in 1960. This increased to
more than 4 lb/bird in 1982 and reached 5 lb/bird in 2000 (figure 11).

Figure 10. Yield of flue-cured tobacco (GA–FL Type 14) by year, 1920—2000 (data plotted at 10-year intervals).

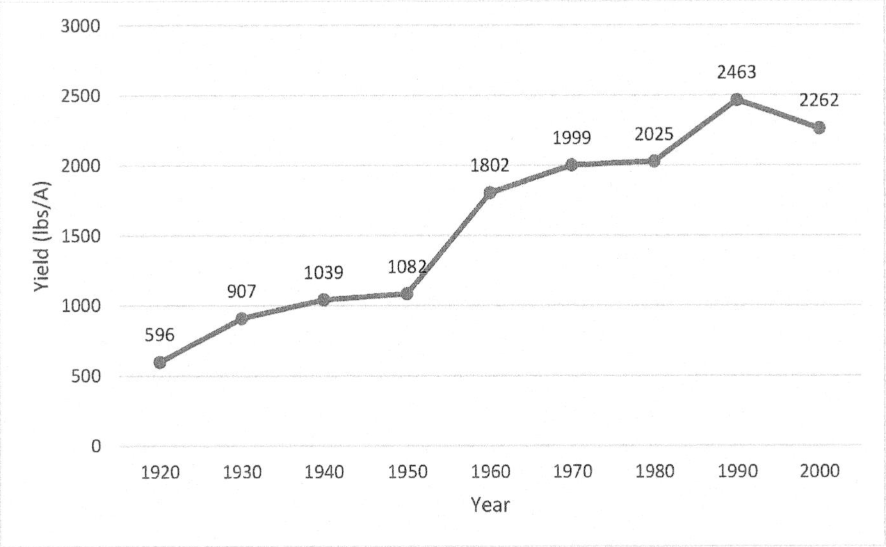

USDA National Agricultural Statistics Service historic databases. See http://www .nass.usda.gov/Quick_Stats/.

Total live weight of birds closely tracked the number of birds slaughtered. In 2000 average live weight was 5 lb/bird with a total production of 41.3 billion pounds. The success in improvement of the feed efficiency in poultry is an impressive accomplishment due to dedicated research in poultry nutrition.

Comparison of Percentage Yield Increases

Percentage increase in yield for selected commodities when comparing first reported yield with 2012 yields is shown in table 1. Of those commodities compared, the percentage increases ranged from above 600% for cotton and 408% for corn to 260% for soybeans, 321% for wheat, 183% for barley, and 268% for flue-cured tobacco. Peanut yields increased 535% during this period.

Figure11. Average live weight of young chickens during the period 1960–2000 (data plotted at 10-year intervals).

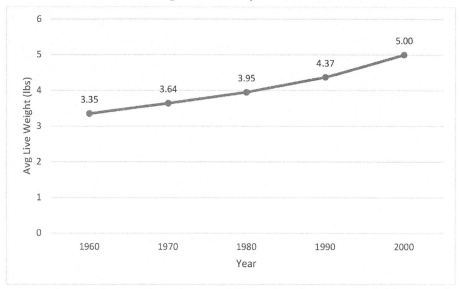

USDA National Agricultural Statistics Service historic databases. See http://www
.nass.usda.gov/Quick_Stats/.

Table 1. Percentage of Increase in Yield for Selected Commodities

Crop	corn	soybean	cotton	peanuts	wheat	barley	tobacco
first reported yield	24.3	11	122	660	11.0	24	488
last reported yield	123.4	39.6	866	4192	46.3	67.9	1798
% of increase	408*	260	610	535	321	183	268

*Because of drought throughout much of the Midwest in 2012, average yield was
unusually low. A more realistic average yield for corn in the United States occurred
in 2010, when it was 152.8 bu/A. When using the latter figure, the percentage of
increase is 528%.

Exports of Agricultural Commodities

Another means of directly measuring agriculture's success and indirectly the importance of agricultural research is to analyze the exports of agricultural commodities. When employing this metric, investments in agricultural research are justified. For example, the United States consistently exports substantial amounts of agricultural commodities. Exports of corn for each year during the past decade have been as high as 45 million metric tons (MMT) but were only 38.4 MMT in 2011–2012 and only 18.3 in 2012–2013. Exports in recent years have been lower primarily because of a drought in the major corn producing regions of the United States, as mentioned earlier. Major exports of other commodities, including wheat, soybeans, rice, and cotton, have also been recorded (table 2).

The United States consistently exports corn grain to more than fifty countries. While some countries receive only modest amounts, Japan, on the other hand, imported in excess of 15 million metric tons of US

Table 2. Selected Farm Products: US Exports 2000–2010 (in Metric Tons, except as Indicated; e.g., 28.9 represents 28,900,000)

Commodity	Unit	Amount			United States as a Percentage of World Exports		
		United States					
exports[1]	million	2000	2005	2010	2000	2005	2010
wheat[2]	million	28.9	27.3	34.7	28.5	23.3	27.8
corn	million	49.3	54.2	48.3	64.2	66.9	53.2
soybeans	million	27.1	25.6	42.2	50.5	40.36	44.1
rice, milled basic	million	2.6	3.7	3.6	10.7	12.3	11.3
cotton	million bales[3]	6.7	17.7	15.5	25.7	39.4	41.9

1. Trade years may vary by commodity.
2. Wheat, corn, and soybean data are for trade year beginning in year shown.
3. Bales of 480 lb. net weight.

Source: USDA, Foreign Agricultural Service, "Production, Supply and Distribution Online," http://apps.fas.usda.gov/psdonline/psdhome.aspx.

corn in 2010 and 2011. Export of soybeans has increased significantly from 27,100,000 metric tons in 2000 to a high of 42,200,000 metric tons in 2010. On the other hand, cotton has experienced the greatest increase in exports during the past decade primarily because of the demise of the textile industry in the United States. The United States exported just slightly more than 6 million bales in 2000 and more than 10 million bales each year since 2003 and 2004. In 2010, more than 15 million bales were exported.

Among the consistently highest export commodities are unmilled wheat, fruit juices, wine, feed grains, feeds and fodder, oil cake and meal, and oil seeds (table 3).

To illustrate this, in 2010, 14,986,000 hectoliters of fruit juices and wine were exported. In that same year exports of unmilled wheat were 27,592,000 metric tons; 57,794,000 metric tons of feed grains; 18,925,000 metric tons of feeds and fodder; 10,010 metric tons of oil

Table 3. Agricultural Exports: Volume by Principal Commodities, 1990 to 2010 (in Thousands; e.g., 7,703 Represents 7,703,000)

Commodity	Unit	1990	2000	2005	2007	2008	2009	2010
fruit juices and wine	hectoliters	7,703	14,356	13,982	14,470	14,871	13,675	14,986
wheat, un-milled	metric tons	27,384	27,568	27,040	32,991	30,021	21,920	27,592
feed grains	metric tons	61,066	54,946	50,865	63,215	59,659	51,388	54,794
feeds and fodders[1]	metric tons	10,974	13,065	11,422	11,823	14,372	14,594	18,925
oilcake and meal	metric tons	5,079	6,462	6,905	8,272	8,405	9,251	10,010
oilseeds	metric tons	15,820	28,017	26,462	31,077	35,011	41,210	43,297

1. Excluding oil meal.

Sources: USDA, Economic Research Service, "Foreign Agricultural Trade of the United States (FATUS)," http://www.ers.usda.gov/data-products/foreign-agricultural-trade-of-the-united-states-(fatus).aspx, and "Global Agricultural Trade System," http://apps.fas.usda.gov/gats/default.aspx. U.S. Census Bureau, Statistical Abstract of the United States, 2012.

cake and meal; and 43,297,000 metric tons of oil seeds (US Census Bureau, *Statistical Abstracts of the United States,* 2012).

Major export markets for US corn include Japan, Mexico, and South Korea, while the top markets for wheat are Nigeria, Japan, and Mexico.

Table 4. U.S. Agricultural Trade: Fiscal Years 1995–2014 (Year Ending September 30)

Billion Dollars			
Year	*Exports*	*Imports*	*Balance*
1995	54.6	29.9	24.7
1996	59.8	32.6	27.2
1997	57.3	35.8	21.5
1998	53.6	37.0	16.6
1999	49.0	37.5	11.5
2000	50.7	38.9	11.9
2001	52.7	39.0	13.7
2002	53.3	41.0	12.3
2003	56.2	45.7	10.5
2004	62.3	52.7	9.6
2005	62.5	57.7	4.8
2006	68.6	64.0	4.6
2007	82.2	70.1	12.1
2008	114.9	79.3	36.6
2009	96.3	73.4	22.9
2010	108.5	79.0	29.5
2011	137.4	94.5	42.9
2012	135.8	103.4	32.4
2013	140.9	103.8	37.1
2014	142.6	110.0	32.6

Reflects forecasts in the February 8, 2013 *(World Agricultural Supply and Demand Estimates report).*

Source: Compiled by USDA using data from US Census Bureau, US Department of Commerce.

China buys more US soybeans than the remainder of the world combined. Mexico, Russia, Hong Kong, and Canada are our major markets for poultry meat.

Agricultural commodities are one of the few bright spots in the US balance of trade. In the past, bulk commodities, including wheat, rice, corn, oilseed, cotton, and tobacco, constituted most of our exports, but in recent years exports of high-value products such as meats, poultry, oilseed meals, vegetable oils, fruits, vegetables, and various beverages have shown steady growth. Over the past two decades the value of agricultural exports has been about $50 billion or greater each year (table 4).

Each year has shown a favorable balance of trade, with a positive balance ranging from $4.6 billion in 2006 to a high of $42.9 billion in 2011.

Elimination of Drudgery

Though enhancing productivity in agriculture is the most visible and a noteworthy benefit of agricultural research, the elimination of drudgery, particularly for those engaged in production and processing, is highly important. For example, the development of herbicides is an excellent example. Removing weeds from crops and in other situations represents drudgery of the highest order. One has only to remove crabgrass from peanuts, common Bermuda grass from corn, or cogon grass from any place to experience and appreciate this problem.

Harvesting research has taken some of the toil out of this step in agricultural production. Until the middle of the twentieth century corn was harvested by hand one ear at the time, with the harvest season often lasting several months. Picking cotton by hand was a tedious, menial, and boring task. Although the cotton gin was invented in the late eighteenth century, the mechanical cotton picker was not developed until the mid-twentieth century. Harvesting of small grains has been evolving since the development of the hand scythe. Then came the reaper and the stationary thresher, and later the combine with a bagger.

Finally, in the mid-twentieth century, combines were complete with a grain tank and a self-unloading auger. These changes represented a steady progression of improvements that eliminated hard labor and improved efficiency. Harvesting of peanuts evolved in a similar pattern. Today the digger-inverter-shaker, accompanied by the peanut combine, takes away much of the drudgery of peanut harvesting.

Harvesting many of our fruits and vegetable crops still requires substantial investments in manual, often referred to as "stoop," labor. However, the development of mechanical harvesters for some fruit and nut crops has been successful. Mechanical harvesting of canning tomatoes has completely revolutionized the tomato industry. Blueberry harvest is at least partially mechanized by using a mechanical shaker. Nonetheless, crops such as figs, peppers, eggplants, squash, and strawberries are harvested exclusively by hand.

Harvest of nuts such as walnuts, almonds, and pecans is routinely done mechanically. Nuts are shaken from trees mechanically, then windrowed and picked up with a mechanical sweeper. The need for manual labor in the harvesting and processing of some crops and foods has resulted in a great migration of workers to this country. Many of these people with basic on-farm skills are willing to do such backbreaking work to survive. Even though there is only modest drudgery associated with animal production, processing food animals represents one of the more undesirable jobs in our society. Consequently, here again immigrants are often willing to do these unpleasant tasks.

Improvement in the Quality of Agricultural Products

Despite the great diversity of the plant and animal kingdoms, about three-fourths of food for human nourishment comes from approximately a dozen plant and a half-dozen animal species. In fact, of the thousands of edible species of plants only about 200 are used to any great extent. About 60% of the calories and proteins derived from plants come from three species: corn, wheat, and rice.

In view of this relatively narrow genetic base for survival of the human species, there certainly should be a strong commitment to improving the quality of agricultural commodities. As more is learned about specific nutritional requirements, plants can be developed to meet each one by using the power of genetic engineering. For example, genetic lines of lettuce with added levels of nutrition equal to exceedingly high levels of folate and vitamin C have been developed.

The Bill and Melinda Gates Foundation has committed considerable resources to improving the nutritional status of several staple foods through genetic improvement. Some of their efforts are enhancing the vitamin A content of sweet potatoes, maize, cassava, rice, and bananas; iron in beans, cassava, and bananas; and zinc in wheat and pearl millet. Cultivars of peanuts and soybeans that have high levels of specific amino acids have been developed.

In recent years a concerted effort has been made to address the problem of peanut allergy. Researchers are working to identify the genes that account for the production of the specific allergens responsible for these allergic responses. Of course, the ultimate goal is to mitigate the effect of the proteins so that peanuts do not cause a severe allergic reaction in humans.

One of the exciting advances in the improvement of food quality through research is the development of golden rice (mentioned in chapter 5). Utilizing techniques of genetic engineering, a variety of rice (*Oryza sativa*) has been developed to biosynthesize beta-carotene in the edible parts of rice (Ye et al., "Engineering the Provitamin A [Beta-Carotene] Biosynthetic Pathway"). Beta-carotene is the precursor of vitamin A. Inadequate levels of vitamin A pose a serious health problem for people in many highly populated countries in parts of Asia, Africa, and Latin America. One study estimates that vitamin A deficiency is responsible for 600,000 deaths each year in children under 5 years of age (Black et al., "Maternal and Child Undernutrition"). Later research led to a new variety called "Golden Rice 2," which produces up to twenty-three times more beta-carotene than the original golden rice (Paine et al., "Improving the Nutritional Value of Golden Rice"). Other studies have shown about a threefold increase in beta-carotene in transgenic tomatoes (Römer et al., "Elevation of the Provitamin A Content").

Agricultural research that enhances the unique nutritional properties of food certainly makes important contributions to quality of life. As we learn more about human nutrition and the relationship between food and health, such studies promise to offer even greater opportunities for accomplishments in agricultural research. Also, it is exciting to speculate on the potential to develop the capacity of plants to synthesize specific medicinal compounds, thereby opening up further opportunities for agricultural experimentation.

Agricultural research has led to countless improvements in other aspects of production. The tomato harvester was a great invention, but it could not reach its full potential until cultivars of tomatoes were developed that could be harvested by such machines. Many varieties of crops were developed to withstand the rigors of shipment and arrive at markets in good shape. Today we can enjoy watermelons, strawberries, and many other crops that without research would be available only in areas near the site of production.

Improvement in Quality of Life

Much of horticultural research in turf and ornamentals has contributed to enhancing our quality of life through improvements in the landscape and the environment. Development of high-quality cultivars of turf grasses for sport and leisure activities is another measure of the success of agricultural studies. Much of the beauty we enjoy throughout our environment is supported or enhanced by agricultural research.

Some aspects of forestry research also contribute to quality of life. For example, studies that focus on multiple uses of forest resources (e.g., hunting, hiking, camping), along with wood production and grazing, protection of minerals, watersheds, and wildlife habitats, all contribute greatly to the quality of life. Also of considerable importance is the simple fact that our forests fix tremendous amounts of carbon, which is of increasing concern.

To address issues of water conservation, some research has led to more drought-tolerant turf grasses and shrubbery for homes and businesses.

Intangible Contributions from Agricultural Research

Agricultural research is responsible for and has made countless intangible contributions that enrich lives. For example, research that led to improvements in the internal structure of cotton has made such a contribution. This wrinkle-free cotton fabric enables people to wear highly desirable cotton clothing without ironing.

Much of the research in food processing provides time-saving approaches to preparation of meals. An excellent example is instant potatoes. It is hard to put a dollar value on such accomplishments, but it is safe to assume that such research has great value. Further processing of commodities results in shorter food-preparation time, thereby allowing the food preparer to engage in other meaningful work.

Another major intangible contribution from agricultural research is success in developing means of controlling (managing) Formosan termites. Without a dedicated research effort led by the USDA Agricultural Research Service and including many state agricultural experiment station and industry scientists, this particular pest would have gotten out of hand, resulting in the loss of older wooden structures in locations such as the French Quarter in New Orleans. These achievements illustrate how the combined agricultural research community, including federal, state, and industry scientists can collaborate to solve seemingly insurmountable problems.

Reducing Costs per Unit of Output

Increasing yield and quality of output are important accomplishments that contribute to the success of agricultural research. However, as pointed out in chapter 8, the most important goal is to reduce the cost per unit of output. Improving cost per unit of output enhances the total factor productivity (TFP). This concept is discussed fully in chapter 8.

Specific Research Accomplishments

To say that agricultural research is a diverse enterprise is an understatement. First, the research enterprise itself ranges from casual observation and selection to highly sophisticated, controlled experiments with treatment effects measured by careful statistical analyses. Next, "who does research?" ranges from farmers to homeowners who make a simple observation and then test their observation or hypothesis, to cooperative extension specialists, who are constantly trying something to solve some particular problem on the farm, to researchers in universities, governments, or private laboratories using sophisticated equipment, procedures, and research approaches.

Identifying specific research accomplishments is further complicated because so many investigators from multiple institutions work on a particular problem for an extended period of time. Indeed, in agricultural research, "eureka" moments are few. In a sense, agricultural research accomplishments are diffuse and do not easily lend themselves to a specific, neatly packaged achievement. Consequently, those who toil in the research field usually derive their satisfaction from knowing that they have made a small contribution to solving problems and issues much larger than those that any one individual could address.

Examples of Specific Accomplishments in Agricultural Research

The accomplishments and successes of agricultural research are far too numerous to enumerate; however, a few examples will illustrate how such efforts touch the lives of all people.

Boll Weevil Eradication

The boll weevil arrived in the United States from Mexico late in the nineteenth century. The beetle has, in the past, been the most destructive insect pest with regard to cotton and among the most destructive agricultural pests in the United States. At one time more than 40% of all insecticides used in agriculture were aimed at controlling the boll weevil. Research in the early twentieth century showed some promise for control by chemicals and cultural practices. Education was an important contributor to this success, leading to the concept of extension and ultimately to the beginning of the Cooperative Extension Service. Over the years much research was directed at controlling the boll weevil. A wide range of pesticides was evaluated in this regard, including calcium arsenate, DDT, toxaphene, aldrin, dieldrin, endrin, heptachlor, malathion, and parathion. Also included in the effort were various cultural practices.

The problem became so severe that Congress supported the creation of the USDA Boll Weevil Research Laboratory. The basis for an eradication program was provided by pioneering research by entomologists published in 1959, outlining the winter dormancy (diapause) behavior of the boll weevil (Brazzel and Newsom, "Diapause in *Anthonomus grandis* Boh"). Other studies showed that implementing control treatment during diapause, coupled with various cultural practices, reduced the overwintering population. The effort got another boost when research identified the boll weevil pheromone, which facilitated trapping and monitoring the boll weevil.

While research showed what was possible, it took the US government, along with state agencies, including agricultural experiment stations and the Cooperative Extension Service, and other groups, including farmers and industry, working cooperatively to bring success to the project. This research effort has been highly effective. Before the boll weevil eradication program, insecticide applications were made at 7–10-day intervals. As many as 15–18 insecticide applications per season were often required to control the weevil. Today, cotton farmers often use only one, two, or even no applications per season to control the boll weevil. Though this research has been a major factor in returning cotton to profitability, an added significant benefit is the release of fewer insecticides into the environment in order to control this pest.

Sweet Clover and Blood Anticoagulants

The relationship between sweet clover hay and the discovery of the anticoagulant dicoumarol by agricultural scientists is an interesting story. In the early years of the twentieth century, a serious disease found in cattle throughout the Midwest was referred to as hemorrhagic sweet clover disease. Animals stricken with this illness after consuming sweet clover hay bled to death. Obviously, this was a terrible sickness. The problem was brought to a head one cold winter day in 1933, when a farmer brought evidence, including a dead heifer and a quantity of blood completely lacking in clotting capacity, to the attention of a veterinarian, who advised the farmer to go to the Wisconsin Agricultural Experiment Station to see what could be learned.

Building on earlier observation and research, scientists at the experiment station noted that improperly cured sweet clover hay could induce hemorrhagic disease in cattle. Evidence pointed to a deficiency in prothrombin as the cause of this problem. For the next five years, Karl Paul Link, along with his students and staff, worked to isolate and characterize the hemorrhagic agent. This research effort yielded a small quantity of the recrystalized anticoagulant. A next step was to identify the structure, which was found to be 3 3'-methylenebis-(4 hydroxycoumarin), later named dicoumarol.

Collaboration with medical personnel showed the potential of dicoumarol to control the clotting of blood in humans. As a result, it became an effective anticoagulant. While dicoumarol was first used in human medicine, it turned out that the synthetic analog, warfarin, worked better in humans, and warfarin is the form now used in human medicine under the brand name Coumadin. Developing an understanding of the sweet clover disease in cattle ultimately led to other results that provided great benefit for people.

Mass Production of Penicillin

Alexander Fleming, a Scottish bacteriologist, discovered the great therapeutic value of penicillin isolated from mold. Unfortunately, he was not able to mass-produce the antibiotic. Then, in July of 1941, two British scientists visited the United States and brought with them the mold, which had been converted into a stable brown powder. On arriving in the United States, they were directed to the USDA's Northern Laboratory, now a component of today's USDA Agricultural Research Service, in Peoria, Illinois. The scientists in Peoria immediately started their cultures of penicillium. By November 26, 1941 (just days before Pearl Harbor), Andrew J. Moyer, the lab's expert on the nutrition of molds, had succeeded, with the assistance of Dr. Norman Heatley, in increasing the previous yields of penicillin ten times. They added milk sugar to the medium, and again the penicillium mold doubled. Moyer also figured out how to use deep vats to grow the cultures. So encouraging were the results that four US drug companies agreed to attempt large-scale production of penicillin. Nevertheless, by March 1942, they had produced enough of the drug to treat only a single case. Then the Peoria researchers made yet another breakthrough. Searching for a superior strain of penicillium, they found it on a moldy cantaloupe in a Peoria garbage can. When the new strain was made available to drug companies, production skyrocketed. Thanks to the combined efforts of many people, penicillin was available in quantity to treat Allied soldiers wounded on

D-Day, as well as the world's people ever after.
Wrinkle-Free Cotton Fabric

While corn may be considered our national crop, cotton is an integral part of US history. Cotton, which is indigenous to the Americas, Africa, and India, was independently domesticated in both the old and the new worlds. The earliest record of the use of cotton was in the Old World about 7,000 years ago. Cotton fiber is soft and pliable and, when woven into cloth, is highly desirable for clothing. Invention of the cotton gin in the eighteenth century and the cotton picker in the twentieth century, along with other scientific developments, made cotton available to the masses. However, in the 1950s and 1960s, demand for cotton fabrics began losing ground to easy-care synthetics. Using monofunctional agents to modify cotton, Ruth Benerito, a USDA-ARS scientist, demonstrated a new mechanism for imparting wrinkle resistance that made cotton fabrics as easy to care for as synthetics. Benerito's work led to new products and new procedures in the fabric, wood, and resin industries. She showed that radio-frequency cold plasmas can replace the polluting sodium hydroxide in cleaning cotton. Benerito also discovered a method of treating cotton fibers so that the cellulose molecules were chemically joined. This resulted in what is known as "cross-linking," making the cotton resistant to wrinkling. Her original methods of analyzing cellulose products have benefited a variety of industries in the United States and around the world. Everyone who enjoys wearing the natural fiber cotton, which is not only comfortable but also looks nice, owes a debt of gratitude to agricultural research and the scientist who made it possible.

Great Expectations from Expected Progeny Differences

Selection of parents to produce the next generation of offspring is key to improving the genetics of a herd of cattle or other livestock. Until the development of the concept of expected progeny differences (EPD) for selected traits, about the only way of evaluating potential parents was by selection based solely on phenotype or laborious progeny tests. An animal scientist at Iowa State University, Richard Willham, often called the "father" of EPDs, spent much of his career developing a more effective method of estimating the genetic value of an animal as a parent. Dissimilar EPDs in two individuals of the same breed predict differences in the performance of their future offspring when mated to animals of the same average genetic merit.

Calculation of EPDs includes information on individual performance, ancestors, collateral relatives, and progeny and involves complex statistical equations and models. Such performance records are adjusted for age, sex, and other confounding factors. Moreover, EPDs are particularly useful in comparing individuals for specific traits of interest, such as growth, calving ease, birth weight, milk production, yearling weight, weaning weight, yearling height, rib-eye area, percent retail cut, docility, scrotal circumference, gestation length, stag ability, carcass weight, fat thickness, and marbling. Any discussion of animal breeding and genetics includes a discussion of EPDs. Willham was a true visionary as well as an outstanding scientist who not only developed the concept of EPDs but also worked to incorporate the concept into modern animal production, which revolutionized the industry.

Putting the Screws to the Screwworm

The screwworm is the only insect that is known to consume the living flesh of warm-blooded animals. Existing in tropical and subtropical environments, the screwworm causes massive losses in food and companion animals as well as wildlife and occasionally humans. Living on a farm in the South in the 1940s and 1950s, I witnessed firsthand the destruction caused by this pest. Monetary losses were substantial, but the pain and suffering of animals were even greater concerns. At this time few effective treatments existed. Even then, treatment was usually applied after considerable flesh had been destroyed.

Edward F. Knipling and Raymond C. Bushland, USDA agricultural research scientists, initiated research on improvement of screwworm control measures. However, they soon realized that eliminating the insect was a better approach than treating an individual after an infestation had eaten away much of the flesh. They happened to read a book that discussed the use of radiation to sterilize the screwworm.

This knowledge, coupled with the fact that the female screwworm fly generally mates only once in a lifetime and retains the sperm for fertilization of all subsequent egg batches, enabled Knipling and his colleagues to put two and two together. After developing quantities of sterile screwworm males and releasing them on test islands, they started an eradication program in Florida. Success there enabled them to repeat their success in the remainder of the United States. Results of this research had great economic value, but the contribution to the humane concern for animals is incalculable.

Summary

These few examples are a very small sample of agricultural research accomplishments that have made life better for everyone. There are many summaries of agricultural studies. In fact, most state agricultural experiment stations, USDA agencies and laboratories, and other agricultural research organizations and institutions have such documents. Successes by industry in agricultural research are usually packaged and offered for sale. One document that is helpful in appreciating the breadth of the agricultural research enterprise is the USDA-ARS *Timeline of 144 Years of Ag Research*. This splendid summary is available at http://www.ars.usda.gov/IS/timeline/comp .htm.

7
Beneficiaries of Agricultural Research

In every truth, the beneficiaries of a system cannot be expected to destroy it.

—A. PHILIP RANDOLPH

"Who benefits from agricultural research" is both a relevant and an important question. As the general concept of research was evolving, this was not a particularly important or even a relevant consideration. In fact, most of the people in the world were not interested in many of the findings of early investigators. Much of the early research in astronomy, mathematics, medicine, and so on was for personal satisfaction, although medical research ultimately led to benefits to individuals. Galileo probably conducted his investigations for personal interest and satisfaction. Certainly, his findings did not provide benefits that were recognized by the public at large. Unfortunately, in fact, they brought him into conflict with church doctrine. Consequently, he was placed under house arrest and spent the last 18 years of his life at his country home outside Florence, Italy. Galileo was exonerated by the Roman Catholic Church in 2008, about 400 years after his arrest. One church official is alleged to have described the original ruling against Galileo as a tragic mutual incomprehension (Coyne, "Tragic Mutual Incomprehension"). How true.

As the research process evolved and was later embraced by the agricultural community, the question of "who benefits" became more relevant. Public funding for anything, particularly in a democratic society, calls this into question. Of course, proprietary research benefits the developer or owner of the study. There is little doubt that the beneficiary is a user who is willing to pay the owner of the research, thus providing a profit incentive to undertake the investigation in the first place.

Everyone who partakes of food benefits from agricultural research. Beneficiaries may be categorized in three main groups: farmers, who are a critical but an increasingly smaller group; consumers, who include every person on the planet; and finally, society as a whole. Much research that is done for the "common good" and does not readily lend itself to profit making is usually paid for by public monies. Consequently, the issue of "who benefits" becomes highly relevant, especially when justifying budget requests.

By the very nature of agricultural research, some of its beneficiaries are easily identified. This is particularly true for the applied component of such experiments. Today fewer than 5 million Americans (less than 2%) live on a farm. This is down from more than 22 million who lived on farms in 1880, when the US population was only 50 million. Of the 2.2 million farms in the United States today, fewer than a quarter million account for more than 60% of sales of agricultural products. Those who live on a farm today and are directly engaged in production agriculture represent only a small fraction of the total agricultural and food portfolio. Many others contribute to the overall process.

Farmers

Farmers are often considered the only beneficiary of agricultural research. Certainly, they are beneficiaries, but they are not the only segment of the population that does. Farmers reap the rewards of this work in multiple ways. Some research solves problems that make the production of a crop or an animal enterprise possible. Basic soil fertility research falls into this category. Developing the most efficient level of fertility gives farmers advantages in improving production efficiency. Of course, crop cultivars are the result of experiments and are extremely important. The farmer also benefits from research that focuses on minimizing the impact of plant pests, including weeds, insects, nematodes, diseases, and various other factors that affect yields and quality.

Research has contributed in a myriad of ways to improve cultural practices, including cultivation and fertilization regimes, optimum time of planting and harvesting, crop spacing and rotation, and many more. Harvesting and processing crops have received a great deal of research emphasis and have enjoyed successes in some areas. Many crops (e.g., all grains, peanuts, potatoes, many nut crops) are harvested completely by mechanical devices. Even fruits such as grapes and blueberries, which present the greatest challenge, have seen some success in mechanical harvesting.

The farmer also derives many benefits from food animal studies. Genetic improvement has been noteworthy in all food-animal categories, significant improvements have occurred in how animals are produced. Humane treatment of food animals is an increasing concern in our society. In view of this, much research effort is directed toward meeting this concern.

Another focus of research that benefits the farmer is the development of new agricultural products. This is such an important area of research that the USDA created four "utilization" laboratories whose focus was initially to develop new products from agricultural commodities. Today, these laboratories have a much broader mission. These USDA Regional Research Centers, located in Albany, CA; New Orleans, LA; Wyndmoor, PA; and Peoria, IL, have proven to be a great investment in agricultural research.

Enhancing productivity is a worthwhile goal of agricultural research. In addition, one of the most important research benefits, as far as the farmer is concerned, is the elimination of some of the drudgery of farming. As pointed out in chapter 4, plowing and cultivating with animal power requires the farmer to walk countless miles, often under difficult conditions. For example, dry conditions create problems with dust, which saturates the nostrils and infiltrates the lungs.

Controlling pests such as weeds and insects has also taken much of the hard work out of farming. For example, I recall as a young boy hoeing common Bermuda grass in corn. Quite an impossible task. Also, trying to control insects with ineffective insecticides was frustrating. Harvesting many crops without mechanization was not only backbreaking but also highly inefficient. Agricultural research has eliminated much of the toil and has the potential to eliminate even more. Treating animals that had been attacked by screwworm was almost unbearable from a humane perspective. Once in place, the screwworm eats away flesh to bone tissue. Research efforts that led to the eradication of this scourge of food and companion animals in the United States is one of the greatest accomplishments in the history of agricultural research.

Certainly, there is no argument that the farmer is one of the primary beneficiaries of agricultural research, and the success of the American farmer is not questioned. Today, less than 2% of the population of this country is engaged in agricultural production. This figure is, in some way, a bit misleading because it does not include many others involved in the seed, pesticide, fertilizer, banking, communication, transportation, and other industries that constitute the total agricultural effort. Each of these contributes in some way to the phenomenal success of agriculture in the United States. Though some may argue about the precise number engaged in production agriculture, the simple fact is that the American farmer is the most successful agriculturalist on Earth, and much of that success is attributed to agricultural investigation.

It would be comforting to say all farmers benefit from the results of research. Unfortunately, this is not necessarily true. A range of benefits occurs for different farmers. Those who take the risk of adopting new technology stand to gain the most from such innovations. If the new technology is substantial, greater productivity could lead to lower prices per unit of product. Of course, the early adopters can still be competitive because of the increased productivity. The nonadopter of new technology would be in a bind with no increases in productivity and lower prices.

In some instances the introduction of new technology can change the face of agriculture. For example, the development of controlled atmospheric storage of apples enabled some of the prime apple-growing regions of the United States (e.g., Washington State) to greatly expand production. With this innovation, they were able to supply the market year-round.[1] When such storage technology was developed for Vidalia onions, the industry was changed almost overnight from a small local business to a major commodity.

As research strengthens agricultural productivity, farmers certainly benefit, but the added advantage is the total economy of the country. By any metric, our country excels at agriculture. For the well-being of our country, strengthening agriculture just makes sense. Agricultural research is one of the primary means of accomplishing that goal. It is like betting on a team that has won every game of the season. I'd like to bet on that team for the championship!

Consumers

The true and ultimate beneficiary of agricultural research is the consumer. There continues to be a decline in the percentage of disposable income that individuals spend on food in some countries. In the United States around 1900, about 40% of individual income was spent on food. During the early to mid-twentieth century, the percentage fell to 20–25% and dropped even more—to about 15%—in 1960. Today the figure is around 10% of disposable income expended on food in the United States. This is in sharp contrast to countries where disposable income needed for food can range from 35 to 50%.

The decline in disposable income spent in the United States for food results partly from advances in agricultural technology as well as the general growth of income. The current disparities among countries in disposable income spent on food can be attributed to differences in national income levels and in agricultural productivity.

When considering benefits for consumers, two types are evident: one from the development and introduction of new food products and the second from reduction in the price of existing food and drink products. When considering the price of food and drink relative to the price of other goods and services that US households consume, the price of food at home has declined 1% per year from 1948 to 1996. This downward trend continued for another decade before unusual short-term events caused substantial instability in food prices. Consumer gain from reduced food prices was chiefly due to agricultural research. Such benefits represent a high rate of return for public investments in agricultural research (Huffman, "Household Production and the Demand for Food").

Everyone benefits from agricultural research! Some reap more rewards than others, and everyone benefits in a unique way. However, if you eat or live in our modern society, you profit in some fashion from the work done by the dedicated scientists who conduct agricultural research. No area of research is more important to humankind.

Even though research in medicine, space, defense, homeland security, transportation, and other areas are valuable to people, the benefits of agricultural research are unquestionably critical to the survival of our species on this planet.

The majority of the world's population enjoys an abundance of nutritious, highly diverse, and relatively safe food. Conversely, for various reasons, significant numbers of people are not so lucky. Walking the aisles of any major grocery store in this and many other countries shows at first hand agriculture's success. Though agricultural research cannot and should not take all of the credit for our abundant food supply, it does deserve a significant portion of our thanks for the successes of food security. Improvements in production, processing, and transport of food are all the result of either proprietary or publicly funded research.

The array of breakfast cereals is an excellent example of the wide range of nutrients and costs. A number of different grains (e.g., corn, wheat, barley, oats, rice, rye) are converted into breakfast cereal. Likewise, numerous different methods of processing are employed. But that is only a start: These processed grains may receive additives such as honey, sugar, cinnamon, various nuts, raisins, dates, or chocolate to make them even more appealing (and some would argue less healthful) to consumers.

Potatoes can be purchased in bulk at a modest cost or further processed in many different ways for consumer convenience. Such further processing, or reprocessing, may create "new" dishes while meeting the consumer's demand for convenience.

Growing up on a farm, when chicken was on the menu, the process started with selecting a pullet or hen for total processing. Now one has the option of buying either fresh processed, frozen, or canned chicken. Other alternatives include literally thousands of processed types of chicken in contrast to a whole chicken with two legs, two wings, and one breast!

A few of the consumer benefits of agricultural research involve frozen foods and the instant mashed potatoes just mentioned. These two contributions from agricultural research were developed by the USDA-ARS utilization laboratories. The Southern Regional Research Center (SRRC), along with many USDA-ARS scientists, several state

agricultural experiment stations, and a number of industries, cooper-
ated in an areawide management effort to seek ways of eliminating or
diminishing the population of the Formosan subterranean termite. As
mentioned in chapter 6, the target area for this groundbreaking large
cooperative research project was the largely wooden French Quarter
in New Orleans. This effort was successful in reducing the populations
of the winged form (the alate) of the termite from about 70% in 1998
to about 42% by 2002, and the number is still lower today. The number of infested structures was reduced substantially. In the United
States, the cost of the structural damage caused by this termite species
is estimated to be in the hundreds of millions of dollars.

Almost every item in the average supermarket or grocery store
bears a mark of agricultural research either by industry or a public
institution. Those individuals with special dietary needs benefit from
agricultural research in a myriad of ways. Foods that are processed
with low sugar, caffeine, fats, or other undesirable components are
helpful to some people. Others benefit from gluten-free foods. On
the other hand, some consumers benefit from foods that have been
fortified with specific components such as calcium and various other
minerals or compounds. Even though such additions or deletions
occur during processing, much research is now directed toward enabling the plant or animal to produce the desired food product during normal growth. This area of development and discovery is still in
its infancy.

With genetic engineering, it will be possible to incorporate many
more desirable traits into food products in the future. This holds
great promise for low-income people, particularly in the developing
parts of the world. One of the more exciting aspects of this approach
is the incorporation of a plant's natural capacity to synthesize medicinal compounds. In my opinion, the discovery of such capacities will
probably be the next major paradigm in agriculture (i.e., the "marry-
ing" of medicine and agriculture).

While food for humans, feed for livestock and companion animals, fiber for our clothes and homes, and new sources of energy are
clearly objects of agricultural study, a fifth dimension involves flowers, shrubs, and trees for our landscape and the environment: yet another benefit. Considerable agricultural research is directed toward

horticultural science and related areas that protect and add beauty to our environment and landscape. New flower varieties and better cultivars of turf grasses do not just appear; rather, they are a result of dedicated agricultural research. Many sports are played on grasses that have resulted from turf research. These aspects of agricultural experimentation clearly benefit those who appreciate beauty and the environment (e.g., homeowners, naturalists, sportsmen and sportswomen). In addition, the hobby gardener benefits from agricultural research in virtually every aspect of gardening. Agricultural investigations have led to the best soil-fertility practices, pest and disease control and culture, and the management of most garden plants.

Society

One could argue that the greatest beneficiary of agricultural research is society at large. How does society differ from the consumer? Society may be considered from a broad holistic perspective, whereas the consumer can be thought of as an individual. Of course, both society and the individual can be considered from other perspectives. As discussed earlier, our civilization was made possible by the success of agriculture. Until a consistent and reliable source of food was available, people spent much of their time engaged in finding nourishment. This was not just the highest priority; it was the only priority. A person who is genuinely hungry has only one priority.

At present, most countries have an abundance of food. Unfortunately, this is not universally the case. In fact, almost a billion people in the world do not receive a caloric intake adequate for a successful life. Many others do not receive adequate protein, fats, and other specific nutrients (Food and Agriculture Organization of the United States, "The State of Food Insecurity"). Indeed, a great disparity in the availability of food exists among nations.

Some benefits from higher agricultural productivity to society are sometimes overlooked. Although intensive agriculture can present environmental challenges, it does make possible the maintenance of more land in a natural state, thereby fostering environmental attributes, including biodiversity, wildlife, and recreation. Intensive agricultural production, coupled with conservation programs, has contributed to considerable reduction in soil erosion.

Another great societal benefit is that in our country agriculture has been so successful that food has been made available to many other nations. Agriculture is one segment of our economy that has consistently shown a positive balance of payments with respect to international trade. This is definitely a societal benefit by helping our overall balance of payments.

It should be the goal of any nation-state to be either self-sufficient in food or successful in other pursuits before it becomes a legitimate nation. Such nations could use income from "other pursuits" to purchase food necessary to sustain themselves. Trade should consist of trading the excess commodities and products of one country for those of another country that the former cannot make or produce.

All nations should strive to develop sufficient food-production capacity. Depending upon their resource base, efforts should focus on areas of food production that utilize each nation's own assets. For example, nations with limited arable land but abundant energy and technical resources should concentrate on intensive production and processing or reprocessing methods. Conversely, those nations with great expanses of arable land should focus on less intensive agriculture. Resources such as water will determine the nature of food-production capacity. This goal translates into a clear signal that all nations should develop a robust agricultural research program.

8
Return on Investment in Agricultural Research

An investment in knowledge pays the best interest.
—BENJAMIN FRANKLIN

Agricultural research has provided great benefit to humankind. As described in chapter 7, humankind, as well as society at large, has benefited in numerous ways. Since these benefits come at a cost, the immediate and relevant questions are, at what cost? and, in particular, at what benefit relative to cost? In other words, what is the rate of return (ROR) on investments in agricultural research?

A discussion of the rate of return on investment for agricultural research must begin with a consideration of agricultural productivity. In the past half century, yields of many agricultural crops have increased dramatically, as noted in chapter 6. The magnitude and nature of increases in yield vary with commodity, region of the country, and other factors. A part of the tremendous improvement in yields of agricultural commodities in recent years can be accounted for by using more fertilizer, irrigation, and various pesticides to control weeds, insects, and other pests more effectively. Also contributing is the use of more machinery per worker employed in farming. Better diets and environmental conditions for livestock and increasing herd stocking rates have also led to improvements in productivity. These are examples of increased intensity of agricultural inputs to raise crop and livestock yields. In addition, agricultural yields have been raised by increasing the efficiency with which these resources are used. Instead of "producing more by using more" (i.e., input intensification), this is like "producing more by using less," which came about as we learned better farming methods and adopted new technological systems of production.

Such developments in new farming practices and technologies that enable more effective and efficient use of inputs to enhance production is a concept called "total factor productivity" (TFP). The USDA's Economic Research Service has developed a careful measure of TFP in US agriculture since 1948, taking into account all of the commodity outputs and resource inputs employed in farm production. Its analysis shows that while the sum total of land, labor, and capital inputs employed in agriculture has hardly changed since 1948, by 2009 the output of crops and livestock had risen about two and a half times (Economic Research Service, *Agricultural Productivity in the U.S.*; Heisey, Wang, and Fuglie, "Public Agricultural Research Spending"). While increases in machinery and chemical use have occurred, they have been offset by reductions in labor, land, and other inputs to keep the overall agricultural resource base roughly constant during that time. Thus, virtually all of the growth in US agriculture during that period can be accounted for by changes in TFP, in other words, by improvements in technology, farming practices, and other factors that have enabled farmers to use their time and resources more efficiently (Fuglie, "Total Factor Productivity"; Fuglie, MacDonald, and Ball, "Productivity Growth in U.S. Agriculture"). The key factor in driving the increase in TFP over the long run has been public investment in agricultural research (Alston et al., *Persistence Pays*; Huffman and Evenson, *Science for Agriculture*; Heisey et al., "Public Agricultural Research Spending).

Figure 12 illustrates how TFP has contributed to agricultural growth in the United States in the past 60 years. Using indexes to measure quantity of total output, inputs, and TFP, the chart shows how much these measures have changed over time relative to the base year (1948, where the index value equals 100). During this period, total crop and livestock output grew by an average of 1.58% per year, and by around 2003 the output index topped 250, meaning that the quantity of outputs produced on farms in the United States was more than two and a half times what it was in 1948 (Heisey et al., "Public Agricultural Research Spending"). Meanwhile, the index of total inputs, while rising slightly until around 1980, then declined, and by 2009 was back to nearly what it was in 1948. In terms of the annual rate of growth between 1948 and 2009, total inputs grew by

Figure 12. US agricultural output, input, and TFP indexes.

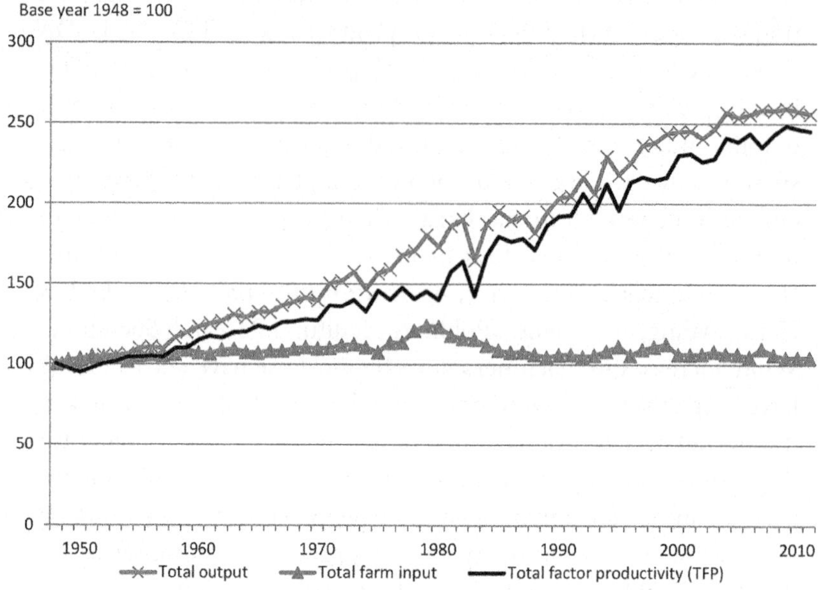

Base year 1948 = 100

—✕—Total output   —▲—Total farm input   ▬▬▬Total factor productivity (TFP)

Source: USDA, Economic Research Service, *Agricultural Productivity in the U.S.*

only 0.06% per year. The TFP grew by 1.52% per year, almost at the same rate as output.

Of course, it is no surprise that a very strong link between growth in agricultural productivity and agricultural research does exist. Indeed, there are numerous constraints on inputs: For instance, the availability of land is constant, fertilizer materials are limited, high-quality water supplies are limited, and feed and seed costs continue to rise. Also, simply applying more inputs can rapidly lead to diminishing returns. For example, each additional unit of fertilizer applied will result in a smaller and smaller increment to crop yield or none at all, with much of the unused fertilizer being lost to the environment.

New technology that enables a crop to absorb more fertilizer and lessens nutrient losses to the environment is needed to offset these diminishing returns. The economic effect of these kinds of technical advances is what is measured in TFP. Therefore, future increases in agriculture's success will be more and more dependent upon TFP, which in turn depends upon the investment in agricultural research.

Response Time for Research

Unfortunately, one of the challenges in this relationship is the lag time between the conducting of agricultural research and significant discoveries of new ideas, innovations, and products, followed by the development, testing, and commercialization of new technologies, and then adoption of these new technologies by farmers. Having spent much of my career working with agricultural researchers and interfacing with legislative bodies, I often was amazed at the lack of appreciation of the nature of the research process by many elected officials. It seemed many state legislators believed that if money was appropriated and a budget passed during the legislative session, then some results should appear soon after the legislature met. Also, they believed that an appropriation of monies ensured research success. Even under the best of circumstances the lag time between research and the initiation and implementation of findings is measured in years, not months, and certainly not weeks. On the other hand, the benefits of research are especially long lived. Today, investments in some areas of research can be expected to contribute to productivity gains at the farm for decades to come.

The problem of lag time between research and the implementation of findings has been addressed by W. E. Huffman and his coworkers, who employed a time frame of 35 years, starting with funding for agricultural research and followed by development and the implementation of the findings (Huffman and Evenson, *Science for Agriculture* and "Do Formula or Competitive Grant Funds"; Huffman and Jin, *Reduced Funding*; Huffman, Norton, and Tweeten, "Investing"). They considered a 5-year research-and-discovery phase, then a slowly increasing period of impacts, followed by a 7-year period of maximum impact and then 20 years of fading impact (figure 13).

Implementation of research results often is slow for multiple reasons. Here are a couple of examples from my personal experiences. A colleague, Dr. Ellis Hauser, USDA-ARS, and I conducted a number of experiments involving peanut row spacing in the early 1970s.

Figure 13. Impacts (w_t) over time of public agricultural research investment in year 0.

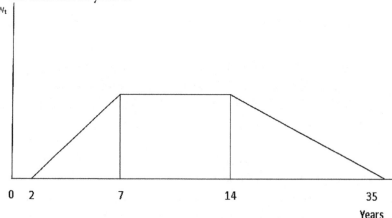

Sources: Huffman and Evenson, "Do Formula or Competitive Grant Funds"; Huffman, Norton, and Tweeten, "Investing in a Better Future."

Our research showed that close-row spacing of peanuts led to both increased yield and better weed control (Hauser and Buchanan, "Influence of Row Spacing"). The results were interesting but were not widely accepted and employed because of the difficulty of adjusting available planters to accommodate close-row spacing. However, in the late 1990s, an innovative twin-row peanut planter came on the market, permitting rows to be planted close together. Research that occurred 30 years earlier was now relevant, and today close-row, twin-row planting of peanuts is common throughout much of the primary peanut-growing region of the United States.

Another example involves the control of early weeds in peanuts. I learned from investigations carried out in the early 1970s that Paraquat, a contact herbicide, was highly effective in controlling a broad spectrum of weeds in peanuts if applied when the weeds were very young. Peanuts were "burned," but they quickly recovered. Attempts to get the company that sold Paraquat to pursue a "use" label for peanuts were unsuccessful. Fast-forward 35 years, when the most widely used, early postemergence applied herbicide (Premerge) for peanuts was taken off the market. Paraquat was quickly identified as a potential replacement.

Huffman and Evenson's illustration of lag time for full implementation of the impact of agricultural research is at best an approximation (Huffman and Evenson, *Science for Agriculture*; Huffman and Jin, *Reduced Funding*). However, in general it illustrates how the research process works. Their model would be much more applicable to some types of research than others. For example, new crop cultivars more closely approximate the model than they do fundamental research, which establishes a basic scientific principle. This illustration does not address the subject of the permanence of research findings, one of the challenges all agricultural research administrations have to consider in working with budget appropriators. Agricultural research is never finished; it is always a work in progress. While a few researchable problems do "stay fixed," many have a relatively long gestational period (Huffman et al., "Investing"). To ensure the success of the agricultural enterprise, it is imperative that all researchable problems be addressed—on an ongoing basis. To make this problem even more challenging, it is usually not possible to predict when a problem will become "unfixed."

Considerable maintenance research is constantly needed to deal with many agricultural pests. Pests are biological organisms that are always evolving and developing mechanisms that enable them to adapt to chemical and/or other treatments. An excellent example is the weed Palmer amaranth (*Amaranthus palemeri*), which has become resistant to the widely used herbicide glyphosate. Many other examples involve other weeds, insects, and plant diseases that have developed resistance to various pesticides. Hence, much maintenance research is needed just to sustain the current productivity of plants and animals.

Determining the rate of return for research in industry is much more straightforward than it is for public-funded research. My personal research career from 1962 to 1980 occurred during a period of great activity in the agrochemical industry. At the time, many companies were synthesizing and testing new compounds for pesticidal properties. It often seemed that a company would initiate a research-and-development program with great fanfare and enthusiasm. If it was able to identify and synthesize a useful compound and was successful in securing a use label, then it was in a position to start making money, and it would be in business and be around for a while.

Private companies often protect their technology through intellectual property such as a patent, which gives them a temporary monopoly during which they can charge a higher price and earn a larger return on their research investment. Once the patent expires, competitors can enter the market and drive down the price and profit margins of the firms that first manufactured the product. Obviously, they were seeking a relatively quick turnaround, or ROR. However, there are many exceptions.

In *Lords of the Harvest*, Daniel Charles describes in considerable detail the beginning of the biotechnology industry as it pertains to agriculture. The Monsanto Company had developed a highly effective broad-spectrum herbicide, Roundup (glyphosate). Though this was good news for killing weeds, it also was very effective in killing crops. A visionary scientist at Monsanto had a breakthrough idea: Why not engineer the desirable plants (crops) to inactivate the herbicide? This was a radically new approach in herbicide development and an unusually challenging research quest involving new and as yet unresearched areas of science. Management at Monsanto became concerned when the ROR on the increasingly costly project did not look promising. The administrators, who were responsible for the company's fiscal solvency, were in a quandary. Should they terminate the project and invest company resources in other ventures? Or should they stay the course with the scientist in this radical new area?

Fortunately, several highly talented scientists were sufficiently persuasive to convince management to continue their support. The success of the "Roundup Ready" gene, which enabled selected crops such as corn, cotton, sugar beets, alfalfa, soybeans, and canola to tolerate glyphosate was revolutionary. The Monsanto biotechnology story is a classic in that people with vision were able to bring about a new paradigm in agriculture. In addition, this story has made multiple contributions to agriculture and agricultural research.

The Monsanto scientists, in defense of their vision, still had to justify the ROR on investment to corporate management. They were successful in devising a means of capturing a profit from biotechnological research by charging a technology fee, much to the chagrin of some farmers. However, this enabled the company to make a profit, which was absolutely necessary for a for-profit company to stay in business. In the end, Monsanto made a profit, and farmers obtained access to technology that they otherwise would not have had.

In a sense, publicly funded research is handicapped compared to industrial research. In industry, if the ROR does not reflect a profit, the plug is often pulled. In contrast, publicly funded research has no equivalent yardstick such as the profit motive. In addition, technology developed by publicly funded research does not usually include a technology fee.[1] This keeps costs to farmers lower but also limits market signals about what farmers want and find to be most valuable in terms of new technology being offered.

There is nothing like dollars of profit to keep score. In fact, most research on ROR has been initiated to provide justification for budget developments or for continued funding to support agricultural research. As pointed out in the previous chapter, our entire society benefits in some way from expenditures in agricultural research. For publicly funded research, the only question remaining is, are the outlays in cost equal to or less than the value of doing the research? This is becoming more and more necessary as money for public research becomes scarce. Consequently, the ROR for any and all agricultural research is becoming an increasingly important consideration.

In comparing ROR for public versus that for private research, it is important to make a distinction between the "social ROR" and the "private ROR." The social ROR includes all economic benefits from research, including those to farmers and consumers. The private ROR would include only the benefits to those who invested in the research, such as what a private seed company can recoup through sales of an improved crop variety it developed and patented. A privately developed seed variety might have some benefits to farmers who adopt it and to consumers who benefit from more abundant food, but such benefits would not be included in a seed company's calculation of its private ROR. Public research does not involve a comparable private ROR since the state experiment station usually does not charge a technology fee for its new technologies (or if it does, it is very low compared to what a private company would charge).

For private research, the private ROR will generally be smaller than the social ROR since the company is unlikely to capture all of

Table 1. Summary Estimates of the Rate of Return (ROR) to US Agricultural Research

Item	Studies (1965–2005)	Mean Estimate of ROR	Median Estimate of ROR
Social rate of returns to public agricultural research	35	53	45
Social rate of returns to private agricultural research	4	45	45

Source: Fuglie and Heisey, *Economic Returns*; Fuglie et al., *Agricultural Research and Development*.

the benefits of the technology it develops. Although we know that the private ROR to private agricultural R&D is less than the social ROR to private R&D, we have little evidence on the social ROR to private R&D. Many studies have measured the social ROR to public R&D. Most likely the social ROR to public R&D will be greater than that to private R&D because the results from public R&D are likely to be more widely used and diffused.[2] Studies have estimated the social ROR to public agricultural research in the United States using statistical models linking historical investments in research to productivity growth. As one example, USDA economists Keith Fuglie and Paul Heisey ("Economic Returns") have compiled an inventory of such studies dating back to 1964 (table 1).

This table summarizes the mean and median estimates of the social ROR to both public and private agricultural research from more than thirty-five such studies, most of which focused on public agricultural research. In all of these studies, the median estimate of the social ROR to research is 45%. The median estimate of the social ROR to private research was also 45%, which is certainly higher than the private ROR earned by the companies that made this investment. A 45% ROR means that a one-time investment of $1 in agricultural research yielded economic benefits averaging about 45¢ every year over the 35-year time frame during which scientific discoveries typically contribute to productivity improvement.

More detailed findings from studies on the ROR to public agricultural research are given in the next table. All of the studies listed in table 2 compared the total cost of federal-state agricultural research in a sector to the value of the increase in US agricultural TFP that this research brought about. While these studies show some variation in the ROR estimates due to differences in methodology and period of time covered, they are remarkably consistent in finding high social returns to public agricultural research. The estimated annual ROR reported in these studies ranged between 22% and 67%. Even at the low end of the estimated ROR (22%), public agricultural research returned far more in economic benefits to society than most other public or private investments of any kind.

Because the studies include total research investment, they include both research successes as well as research failures (or research that has not yet resulted in technologies adopted by farmers) as part of the cost. Research successes are projects that produced new knowledge that has already contributed to higher productivity on the farm. But in the search for solutions to farm problems, scientists may sometimes pursue blind alleys, or perhaps the new knowledge generated by these projects has not yet translated into practical farm solutions. Though these research projects may have uncovered valuable scientific knowledge, the ROR studies treat them as failures because of their limited economic contribution to farm productivity.

In a study that involved both agricultural research and extension, Alston and colleagues (*Persistence Pays*) obtained a cost/benefit ratio of 32 for investing in agricultural research (i.e., for each dollar spent on agricultural research and extension, the return to society was $32). In another study, Jin and Huffman (*Measuring Public Agricultural Research*) reported an updated economic analysis of the contributions of public agricultural research compared to state agricultural productivity for the period 1970 to 2004. They did not find any decline in the ROR to research, which remained at roughly 67% and even over 100% for agricultural extension. Despite the high levels of productivity already achieved in US agriculture and the growing role of the private sector in agricultural research, the returns from using tax dollars to support public agricultural research remain as high as ever.

It should be recognized that the ROR varies greatly with crops and livestock as well as with different enterprises in each of these two groups. This may be due to the nature of the plant, animal, and pest being stud-

Table 2. Estimates from 27 Studies of the Rate of Return (ROR) to Federal-State Investment in Agricultural Research: Studies on the Aggregate Crop-Animal Sector

Study	Authors	Pub. Year	Publication	Period	Coverage	ROR to Research
1	Jin & Huffman	2015	Agric Econ	1970–2004	ag sector	67
2	Wang et al.	2012	book chapter	1980–2004	ag sector	45
3	Plastina & Fulginiti	2012	J Productivity Analysis	1949–1991	ag sector	29
4	Alston, Andersen, James, & Pardey	2011	Am J Ag Econ	1949–2003	ag sector	23
5	Huffman & Evenson	2006	Am J Ag Econ	1970–1999	ag sector	56
6	Gopinath & Roe	2000	Econ of Innov & New Tech	1960–1991	ag sector	37
7	Makki, Thraen, & Tweeten	1999	J Policy Modeling	1930–1990	ag sector	27
8	White	1995	J Ag Appl Econ	1950–1991	ag sector	40
9	Chavas & Cox	1992	Am J Ag Econ	1950–1982	ag sector	28
10	Norton & Ortiz	1992	J Production Agric	1987, state-by-state comparison	ag sector	58
11	Yee	1992	J Ag Econ Res	1931–1985	ag sector	54
12	Braha & Tweeten	1986	technical bulletin, Oklahoma State Univ.	1959–1982	ag sector	47
13	Lyu, White, & Liu	1984	S J Ag Econ	1949–1981	ag sector	66
14	White & Havlicek	1982	Am J Ag Econ	1943–1977	ag sector	22

Study	Authors	Pub. Year	Publication	Period	Coverage	ROR to Research
15	Davis	1979	PhD diss., UMN	1949–1959 1964–1974	ag sector	83 37
16	Knutson & Tweeten	1979	Am J Ag Econ	1949–1972	ag sector	38
17	Lu, Cline, & Quance	1979	technical bulletin, USDA	1939–1972	ag sector	27
18	Bredahl & Peterson	1976	Am J Ag Econ	1937–1942 1947–1957 1957–1962 1967–1972	ag sector	56 51 49 34
19	Cline	1975	PhD diss., Oklahoma State Univ.	1939–1948	ag sector	46
20	Evenson	1968	PhD diss., Univ. of Chicago	1949–1959	ag sector	47
21	Peterson	1967	J Farm Econ	1915–1960	ag sector	23
22	Griliches	1964	Amer Econ Rev	1949–1959	ag sector	33

Studies on Components of the Agricultural Sector

Study	Authors	Pub. Year	Publication	Period	Coverage	ROR to Research
10	Norton & Ortiz	1992	J Production Agric	1987, state-level comp.	beef & swine; dairy; poultry; grain crops; potatoes, cotton & tobacco; vegetables & melons; fruits and nuts	55 95 46 31 34 19 33
23	Haygreen et al.	1986	Forest Prod J	1972–1981	forest products	25

Table 2. (cont.)

Study	Authors	Pub. Year	Publication	Period	Coverage	ROR to Research
24	Bengston	1984	Forest Science	1975, state-by-state comparison	forest products	21
25	Smith, Norton, & Havlicek	1983	J NE Ag Econ	1978, state-by-state comparison	beef & swine; dairy; poultry	22 25 61
26	Schmitz & Seckler	1970	Am J Ag Econ	1958–1969	tomato harvester	42
21	Peterson	1967	J Farm Econ	1915–1960	poultry	23
27	Griliches	1958	J Polit Econ	1940–1955	corn sorghum	38 20

ROR = annual rate of return to research. Most studies assessed the effect of research on productivity over a period of years. "State-by-state comparison" refers to studies that compared research investment and productivity levels among states at a particular time.

Source: Fuglie and Heisey, *Economic Returns*, with new findings from Alston et al., "Economic Returns"; Jin and Huffman, *Reduced Funding*; Heisey, Wang, and Fuglie, "Public Agricultural Research Spending"; Wang et al., "Accounting for the Impact"; Plastina and Fulginiti, "Rates of Return."

ied. Sometimes insects, diseases, and weeds develop resistance to a given treatment, requiring a new remedy or approach. In addition, new cultivars or different cultural practices may require a different approach to keep a problem in check. How well a particular enterprise is adapted to a given area can affect the nature of the ROR.

Even though studies involving different methodologies and other variables show a great deal of variation, the one constant is the remarkably high rate of return per unit of investment in agricultural research. In many areas of government, expenditures of public money do not show an easily identifiable, positive ROR. Clearly, expenditures on agricultural research are for the common good and generally show a positive ROR. All of the billions spent on defense have only one benefit—our safety. Medical research helps to heal us when we are sick. These are appropriate and commendable expenditures of money for the public good. But money spent on agricultural research produces a favorable ROR on investment and sustains us as well as helps ensure the future well-being of our civilization.

As the many economic studies have shown, money invested in agricultural research is expected to yield a return on the order of 20–70%. This is impressive when compared to the average returns of 9% and 12% of the S&P 500 and NASDAQ composite indexes during the same period. In a real sense, agricultural research does not cost anything. Agricultural research pays and pays handsomely. In view of such data, one would think the agricultural research budget would be one of the easiest research budgets for governing bodies to justify. Unfortunately, there is a basic failure to understand or appreciate the importance of agricultural research, particularly the high social ROR on investments.

In one sense, the agricultural research enterprise in the United States is a victim of its own success. The abundance of a highly diverse array of high-quality foods at reasonable costs has led to a sense of complacency about our food supply. Even those who have some degree of food insecurity have for the most part become invisible. Unfortunately, our food system is taken for granted; consequently, most individuals have little or no genuine perception of the need for research.

One of the important challenges facing our society and the entire world is meeting the needs and expectations of the world's growing population, along with increasing expectations of everyone in the years ahead. This does not sound like too much of a challenge until one realizes that agriculture is our most important, reliable, and only source of food and a potential means of achieving energy security.[3] The continuing success of agriculture depends upon many factors, but the most important requirements are new information, new knowledge, and more innovative technology, all of which are gained through research.

9
Challenges in Agricultural Research

The significant problems we face cannot be solved at the same level of thinking we were at when we created them.
—ALBERT EINSTEIN

What makes US agriculture strong? The answer is enlightening. Numerous factors account for the phenomenal success of agriculture in the United States, including a high-quality land base, available water, and favorable climate, geography, and weather, along with good transportation and communication infrastructures. Also important are a favorable business climate with reasonable access to capital and, in general, a rational regulatory climate. Knowledgeable farmers also make an important contribution to successful agriculture.

As the number of those engaged in production agriculture dwindles, the best (i.e., most efficient) are left standing. But the ingredient that has truly made the difference is ready access to information, knowledge, and technology as provided by research and educational systems in both the private and the public sectors. These systems are found in many countries, but the public agricultural research and education system in the United States has been emulated but never fully duplicated anywhere else. This unparalleled commitment to developing new and innovative information, knowledge, and technology has enabled the United States to take advantage of every opportunity to strengthen agricultural productivity. Unfortunately, the job is far from done and becomes more challenging with each passing year.

In a special thirtieth-anniversary lecture commemorating his receiving the Nobel Peace Prize, Norman Borlaug reflected on the breakthroughs in wheat and rice production brought about by agricultural science, generally referred to as the Green Revolution. He pointed out that "only through the dynamic development of a productive agriculture will there be any hope to feed the population growth expected, to alleviate poverty and to improve human health and productivity while reducing political instability." In addition, Dr. Borlaug acknowledged that the Green revolution had won a temporary success in the war against hunger, but the population monster cited by T. R. Malthus is still with us. Before his death in 2010 Dr. Borlaug wrote the following: "In the next 50 years we are going to have to produce more food than we have in the last 10,000 years, and that is a daunting task. I therefore have called for a 'Second Green Revolution'" (preface to Buchanan, Herdt, and Tweeten "Agricultural Productivity Strategies").

The 2014 Borlaug Dialogue International Symposium, held in conjunction with the World Food Prize, featured the theme "The Greatest Challenge in Human History: Can We Feed 9 Billion People by 2050?" One of the lead speakers, Kenneth G. Cassman, the distinguished Dougherty Professor of Agronomy, University of Nebraska–Lincoln, made the following statement: "As we gather during the centennial anniversary of Dr. Borlaug's birth, and as populations continue to rise, the question is whether current trends in food production will adequately provision a stable population of 9–10 billion in a world with a changing climate and provide us with sustainable freshwater resources and acceptable environmental quality. Unfortunately, the answer is *no!*"[1] However, Cassman does offer some hope and encouragement. He points out that we have time to get on track. Among the options we have for changing direction and achieving a positive outcome are enhancing research in the basic and applied sciences that supports innovation in agriculture; more effectively prioritizing research; and facilitating development and technology transfer. This is a wake-up call for action if we are to achieve food security in a future world.

It is abundantly clear that the challenges facing this planet are real. The thesis offered by Thomas Malthus that population, if unchecked, increases geometrically while food increases arithmetically indicates the challenge facing agriculture (Malthus, "An Essay on the Principle of Population"). Frank Elwell argues that Malthus did not attempt to predict what would happen in the future but tried to illustrate the power of geometrical population growth as compared to arithmetical increases in food productivity (Elwell, "Reclaiming Malthus"; *A Commentary; Malthus' Social Theory*). He suggested the following:

- Subsistence severely limits population level.
- When the means of subsistence increases, population increases.
- Population pressures stimulate increases in productivity.
- Increases in productivity stimulate further population growth.
- Because productivity increases cannot maintain the potential rate of population growth, population requires strong checks to keep parity with the carrying capacity.
- Individual cost/benefit decisions regarding sex, work, and children determine the expansion or contraction of population and production.
- Checks will come into operation as population exceeds subsistence levels.
- The nature of these checks will have significant effect on the larger sociocultural system (Malthus points specifically to misery, vice, and poverty).

Even though these principles are relevant, R. L. Thompson ("Proving Malthus Wrong") notes that the world can avoid any potentially undesirable outcomes: "Long ago, British scholar Thomas Malthus predicted that the human population would eventually outgrow its ability to feed itself. However, Malthus has been proven wrong for more than two centuries precisely because he underestimated the power of agricultural research and technology to increase productivity faster than demand. There is no more reason for Malthus to be right in the 21st century than he was in the 19th or 20th—but only if we work to support, not impede, continued agricultural research and adoption of new technologies around the world."

Will Malthus' prediction come true? Not necessarily, but it could. However, several ominous trends on the horizon should be red warning flags: (a) Rate of yield increase for some crops appears to be slowing, (b) funding for agricultural research in many countries is not keeping pace with needs; and (c) global food supplies have begun to fall rela-

tive to demand, along with increasing prices for many basic commodities. Supply and demand have grown closer together. Increased demand
alone provides a sound justification for strengthening agriculture and,
in turn, agricultural research. As world population increases, we have
less room for error in the event of a cataclysmic event.

Fortunately, there are indications that as countries become more
prosperous, birth rates begin to decline. This has occurred in Japan
and some European countries (United Nations, Population Division,
World Population Prospects). If this observation holds for some still-
developing countries, we have reason for a small ray of optimism.
The crucial question for us today is, how can we best meet the challenges that lie in the future?

The medical profession has achieved considerable success in con-
traception research. Studies on this component of the relationship are
outside the purview of agricultural research. On the other hand, almost all aspects of the arithmetical component are driven by agricultural research. Much of agricultural research, however, has worked in
opposition to the geometrical component by reducing the possibility
of famine and its impact on population growth. Each new paradigm
in agriculture described in chapters 4 and 5 have extended the success
and made possible enhanced levels of food production. As Malthus
points out in his 1798 essay, "That population does invariably increase where there are the means of subsistence, the history of every
people that have ever existed will abundantly prove" (Malthus, "An
Essay on the Principle of Population"). Even though the accomplishments in agricultural research are impressive (see chapter 6), the best
is yet to come. This is contingent upon meeting the challenges that
cry out for enhanced research. For the purpose here, the challenge
will be considered from two perspectives, short term (incremental)
and long term (paradigm changing).

Incremental Challenges

Every agricultural commodity that is currently produced has benefited from research that added value in some way. In fact, much of agricultural research is designed and carried out with the idea of making a
positive incremental improvement. Fertility research is often directed at

adding a bushel or two of increased yield with a minimum of enhanced inputs. Another objective is more effective mitigation of various pests. Still other research adjusts cultural practices to minimize input while maintaining or slightly enhancing productivity. Most short-term incremental research does not get to the level of being newsworthy, yet agriculture has greatly benefited from this slow, methodical, and deliberate progression. Indeed, as mentioned in chapter 6, in agricultural research, "eureka" moments are few.

Another aspect of short-term studies is defensive agricultural research. Much pesticide-related research is designed just to protect what we already have. Consequently, it is research done in a defensive mode. An excellent example of such experiments is illustrated by colony collapse disorder (CCD), which began emerging in honeybee colonies in 2006.

Honeybees play a vital role in the food-production system by serving as pollinators for many of our food crops. An estimated 71 of the 100 crops that provide 90% of human food require pollination by honeybees (Woteki, "Road to Pollinator Health"). The approximate value of those crops is as much as $200 billion annually. Loss of foods such as almonds, cucumbers, watermelons, cherries, blueberries, apples, and many vegetables would have a major impact on our diet.

Colony collapse was first noticed by a beekeeper who observed that in the spring some of his hives were empty. In CCD the worker bees leave the hive and do not return, and the resulting loss is devastating to the hive. Losses of hives have averaged 30% or more each year since 2006. Honeybees have plenty of problems (e.g., varroa mites, tracheal mites, a couple species of microsporidia). When it was shown that these were not the cause of CCD, other causative agents, including emerging pathogens, environmental chemicals or toxins, and stressful agricultural practices, were suggested. In addition to these, some researchers suggested the possibility of an insecticidal gene in plants, radiation from cell phones, or stress itself.

Since the problem emerged and was defined in 2006, a great deal of research has been carried out in the identification of the contributory causes of CCD. Although much has been learned, the precise cause of CCD remains unknown. As with all serious problems in agriculture, agricultural scientists are doubling down on their effort to solve the problem. Both the USDA's Agricultural Research Service and the National Institute of Food and Agriculture, working with the nation's land-grant universities, have made solving the CCD problem a high priority (Pettis and Delaplane, "Coordinated Responses to Honey Bee Decline").

Another excellent example of critically important defensive research is the search for control measures for citrus greening. This disease is also known by other names such as Huanglongbing (HLB) and yellow dragon disease. Even though HLB starts slowly, once a citrus tree is infected, there is no cure. Citrus greening is spread by an insect, the Asian citrus psyllid. This disease truly puts America's citrus industry at risk if control measures are not developed. Major research efforts with promising results are under way to solve this problem.

Further illustrating the importance of defensive agricultural research is the development of resistance to Ug99, a wheat stem rust caused by *Puccinia graminis*. This rust was first reported in Uganda in 1999 and has already appeared in a number of countries in Africa and the Middle East. Several races of the Ug99 rust have overcome most of the currently widely used resistance genes. Soon after the emergence of Ug99, agricultural scientists immediately initiated research to seek new resistant genes. One article reports that the Sr35 gene from *Triticum monococcum* is a coiled-coil, nucleotide-binding, leucine-rich repeat gene that confers near immunity to Ug99 and related races (Saintenac et al., "Identification of Wheat Gene Sr35"). Such efforts reflect the importance of defensive agricultural research. Many other examples in agriculture demonstrate that defensive research is critical for a secure food future.

Much short-term research is highly applied. This embraces practical aspects of agriculture such as bull and heifer evaluation in cattle and boar testing in swine. Variety testing of crops is highly popular, as is soil-test calibration. Other applied research involves educating farmers about various cultural practices and crop and livestock production systems, soil-test calibration, evaluation of various pesticides, and so on. Oftentimes, extension scientists work closely with agricultural research scientists in addressing short-term, applied problems in agriculture.

Much of plant and animal breeding is focused on making minor improvements in some desirable quality. The astute researcher is always on the lookout for a unique yet desirable trait that would make a cultivar more useful. For example, a selection of lawn grass that has an improved degree of shade tolerance would be worthy of releasing as a new cultivar.

Most agricultural scientists spend their lives working on incremental improvements for agriculture. Taken individually, many of these would not be noteworthy; however, collectively they can bring about remarkable improvements. One aspect of this never-ending saga is that scientists keep upping the expectation. For example, a plant breeder develops a new variety of peanuts that will produce 50 lb/A more than the current best cultivar on the market. Before the new variety is even on the market, the breeder is already focused on a newer variety that will produce even more than the recently released cultivar and also has a slight degree of resistance to a common peanut disease.

Moreover, unexpected occurrences such as excessive rain or cold change the production dynamics. Also, there are expected occurrences. For example, let's say that the fertility requirements are well established for currently used varieties of a crop. But then the plant breeder introduces a new high-yielding variety that requires different nutritional requirements. Or consider a situation in which irrigation practices are well established for a given crop, but then diesel fuel or electric costs escalate such that water becomes more expensive, or perhaps water is no longer even available. Such developments require another look at best irrigation practices.

Social Sciences and Economics

Sometimes overlooked regarding their importance to agriculture's success are the social sciences and economics. With the dramatic growth in human population and the increasing pace of globalization, it is rapidly becoming more important to be successful in working across national, ethnic, and religious lines. In order to do this, we must become more aware of and sensitive to cultures beyond our own. Collaboration on common issues and problems certainly involves basic science, but implementing solutions requires the human element. Simple matters such as how we produce and process food are considered differently by different cultures. While some cultures accept genetically modified foods, others do not. On some complex issues such as global climate change, a range of acceptance of the science can be found. This, of course, makes finding a solution and implementing it more problematic.

The social question is of paramount importance to trade. For the general prosperity of the planet, production of goods should be done by those who are the most efficient. Trade should consist of those goods that are produced beyond those needed for internal consumption. Moreover, the implementation of new technology brought about by agricultural research certainly requires a human element (i.e., the social sciences) and a sound economic basis to ensure full success. Consequently, the social sciences and economics are vital parts of the challenges that must be met to ensure total success for agriculture.

Failure to appreciate and understand the importance of economics to agriculture can only lead to an unsatisfactory outcome. Every aspect of agriculture requires a sound economic basis for a favorable conclusion. Some of the economic considerations that must be addressed include an assessment of need, which entails an assessment of potential markets. Economic consideration should focus on meeting those needs at an acceptable price. While simple in theory, these matters become exceedingly complex in any agricultural enterprise.

Long-Term (Grand) Challenges

Another category is long-term research, which tends to be offensive in that it pushes the envelope for higher yield, better quality, or some other highly desirable traits. We may think of these as grand challenges. Solving them could lead to a new agricultural paradigm. Of course, development comes in various magnitudes—some bringing about small, incremental impacts, some creating truly paradigm changes, and all levels between these extremes.

Grand Challenges

The following examples are some of the grand challenges in agricultural research that, if solved, would bring about a new agricultural paradigm. In contrast to incremental changes each of these is a particularly challenging scientific problem that will require a major, dedicated research effort to solve. But this is only the beginning. Other important research challenges, if successfully met, would also enhance agricultural productivity. It would be productive for agricultural research leadership to convene an international conference of outstanding agricultural and other related scientists to identify additional relevant grand challenges in agricultural research that could or should be addressed.

Grand Challenge Case Studies

Example 1: Improving Soil Quality

The land base of the planet is essentially fixed. Only about a tenth of Earth's land base is suitable for crop production. Additional extensive areas are more suitable for pasture, range, and forestry. While some new land area is created by volcanic action, it is generally not acceptable for crop production—at least it will not be for a few thousand years.

We have abundant evidence that improving the soil quality strengthens agricultural productivity. Failure to protect the soil can lead to disastrous results, as evidenced by the Dust Bowl in the Great Plains in the early twentieth century. Poor farming practices, coupled with unusually dry conditions, resulted in substantial soil loss, which caused a mass migration of people out of the affected region. The Dust Bowl in the Great Plains is only one example of this phenomenon worldwide. Loss of soil as a result of erosion led to the impoverishment of many areas of North Africa and much of the Middle East. According to the ancient record, North Africa was thought of as the granary of Rome. Lebanon once had cedar forests that covered nearly 2,000 square miles (Lowdermilk, "Conquest of the Land").

Food comes from the earth. The land with its waters gives us nourishment. The earth rewards richly the knowing and diligent, but punishes inexorably the ignorant and slothful.
—WALTER C. LOWDERMILK

Is it possible to develop management systems that intrinsically enhance soil quality by increasing levels of soil organic matter and hence its capacity to retain the available water and nutrients? For example, the *terra preta* soils of Brazil, found in a region dominated by highly weathered and infertile oxisol soils, are particularly fertile. Their patchy distribution on high terraces above the Amazon River and its tributaries and the ubiquitous presence of pottery shards buried deep within the *terra preta* soil profiles, attest to their anthropogenic origin. Some attribute this phenomenon to ash fall from volcanoes or sedimentation in lakes; however, the most likely scenario is that they were deliberately made through slash and char agricultural practices, which incorporated large quantities of charcoal (biochar) and manure into the otherwise infertile native oxisols.[2] This probably occurred over hundreds, if not thousands, of years. Biochar, which is a by-product of the process of pyrolysis, has great appeal because of its potential as a soil amendment.[3]

Within a modern context the question remains whether the biochar coproduct of bioenergy production can be used effectively as a soil amendment to enhance soil quality and recycle nutrients. Pyrolysis and gasification are technologies for transforming biomass into energy products such as syngas and bio-oil, which may be burned directly to produce heat or refined to produce transportation fuels and other products. Both of these systems produce a biochar coproduct, which has been proposed as a soil amendment. The carbon in biochar is highly stable in soil environments, making biochar applications an effective means of sequestering carbon. This is because the carbon in biochar is not (or is only slightly) biologically available.

Thus the question remains whether and how biochar applications can be combined with other practices to raise the levels of organic carbon in the soil and enhance soil biological activity, nutrient cycling, and the productivity of agricultural soils. Biochar applications may be particularly effective in improving our sandy and otherwise poorer soils. Research is currently under way on this exciting potential for systematically improving soil quality, which could be integrated into modern agricultural production practices.

Development of soil conservation efforts led by the US government and supported at the grassroots level brought about dramatic improvements in protecting and improving the soil quality in this country. Conservation efforts embraced a number of enlightened approaches, including cover crops, terraces, grassed waterways, and tree belts, but the most important was the development of an awareness of the importance of conservation and soil improvement. Great potential also exists for soil microbe–plant interaction that would lead to soil quality improvement. The benefits would have multiple effects on crop productivity.

Other opportunities to develop new cover-crop systems could produce yields comparable to those of conventional management but provide the benefits of cover crops, including protection from erosion, reduced nutrient leaching, and the building of soil organic matter.

Example 2: Improving Energy Efficiency of Agriculture

Agriculture is both a potential producer and also a great consumer of energy. One grand challenge of agricultural research is to further improve the energy efficiency of all aspects of agriculture, including production, transportation, and processing. Though the productivity of agriculture in the United States is already impressive, it is also a great consumer of energy. Much of the phenomenal global increases in agricultural productivity, particularly the Green revolution, were based on increased inputs. However, in just a 30-year period, 1950–1980, grain production in the United States doubled with little increase in energy consumption.

In recent decades considerable research has been directed toward improving energy efficiency. For example, the use of reduced tillage has enabled substantial energy savings. More efficient use of fertilizer has also resulted in greater production efficiency. Food-related energy use as a share of the national energy budget grew from 14.4% in 2002 to an estimated 15.7% in 2007. Production of agricultural products accounts for only a portion of the energy use in the food system (Canning et al., "Energy Use in the U.S. Food System"). Direct energy use in agricultural production represented slightly more than 1% of the total energy consumption in the United States in 2002 (Schnepf, "Energy Use in Agriculture"). Direct energy consumption is primarily for fuel (gasoline, diesel fuel, and natural gas) and electricity. Indirect energy consumption is accounted for primarily in the form of fertilizer and pesticides. Direct energy use by agriculture since the late 1970s has declined by 26%, and indirect energy use by 31% (ibid.). Focused research could make further improvements in energy use in these areas.

Some aspects of energy consumption in food systems are accounted for in other sectors. Transportation costs can be substantial, for instance. Moving vegetables from the fields of California or south Florida to the East Coast requires a great expenditure of fuel. Hauling watermelons from Mexico is not cheap. Of course, flying blueberries from Chile makes off-season blueberries expensive.

Example 3: Introducing Nitrogen Fixation in Nonlegumes

Nitrogen is a critical plant nutrient and is necessary for normal plant growth. It is the first of the three most critical plant nutrients. Along with phosphorus (P) and potassium (K), nitrogen (N) makes up the "big three"—NPK. Availability of nitrogen fertilizer is a key requirement for maximum agricultural productivity. Even though nitrogen constitutes more than 78% of our atmosphere, it is expensive when converted to a form plants can use. A number of major crops (legumes) have the capacity to fix atmospheric nitrogen in a form plants can use because of a symbiotic relationship with bacteria.

Some major crops such as soybeans and alfalfa have this capacity to fix atmospheric nitrogen in a form that supports growth of the plant and, on decay, returns it to the soil profile for future crops. The symbiotic relationship of nitrogen-fixing bacteria in legumes, which evolved over long periods of time, is a complex process; however, finding a way to imbue nonlegumes with the capacity to fix their own nitrogen would greatly stimulate productivity. Perhaps other N-fixation mechanisms could be genetically incorporated into new varieties of plants. A number of scientists are working on this challenge; however, full success has so far failed to become a reality. On the other hand, many promising leads have been reported. If the cost of nitrogen fertilizer could be zeroed out for the world's grain crops, then such an innovation would truly be a new paradigm for agriculture.

Example 4: Enabling C$_3$ Plants to Utilize the C$_4$ Photosynthetic Pathway

Photosynthesis is critical for our survival on this planet. Plants have made use of this process for millions of years, and some have made the photosynthetic process more efficient.

There are three basic types of photosynthesis: C$_3$, C$_4$, and crassulacean acid metabolism (CAM). In the C$_3$ type, the first compound resulting from photosynthesis has three carbons (3 phosphoglycerate), hence C$_3$ photosynthesis. In the C$_4$ type, the first compound is a four-carbon compound (oxaloacetate), hence C$_4$ photosynthesis (Slack and Hatch, "Comparative Studies"). In addition, C$_4$ plants, such as corn, sorghum, sugarcane (Hatch and Slack, "Photosynthesis"), and Bermuda grass, are much more efficient at fixing carbon (producing more biomass such as grain, straw, or root mass) than C$_3$ plants are.

The unique structure of C$_4$ plants enables them to divide the reactions of photosynthesis between two types of cells, greatly decreasing photorespiration, a process whereby fixed carbon is released to the atmosphere. Given that some of the most important crop plants, such as wheat, soybeans, and rice, are C$_3$, the capacity to convert C$_3$ plants into C$_4$ plants holds great promise for increasing agricultural productivity. Plants with CAM photosynthesis can essentially become inactive under highly adverse growing conditions. In this type of photosynthesis, CO$_2$ is stored in the form of an acid before use in photosynthesis.

Example 5: Developing Processes for More Efficient Conversion of Cellulose, Hemicellulose, and Lignocellulose to a More Usable Energy Source

Addressing the energy security challenge will require multiple approaches and the use of all available means of generating and conserving energy. Clearly, capturing the sun's energy through biomass is but one of the available approaches. However, it is an extremely important one because it can yield a transportation fuel. While conversion of sugarcane to ethanol and hydrolyzing the starch stored in corn grain and then fermenting the sugar to ethanol has met with considerable success, the net yield of fuel (ethanol) leaves something to be desired. Of course, it would be more desirable to use nonfood biomass for energy production. The true breakthrough will come when we are able to effectively and economically convert cellulose, hemicellulose, and lignocellulose into ethanol or some other suitable fuel or bioproduct. Fortunately, considerable research is focusing on doing just that.

Considerable research is under way to use advanced solid-state fermentation (ASSF) technology to develop a more cost-efficient conversion of the fermentable sugars in sweet sorghum into ethanol. Efforts are currently under way to scale up the continuous solid-state fermentation and distillation process. Results show that fermentation time of the solid substrate is less than 30 hours and that the ethanol yield reaches 90.5% of the theoretical yield (Li et al., "Demonstration Study of Ethanol Production").

Example 6: Enhancing Water Efficiency of Crop Species

As population continues to grow, demand for all resources, particularly water, will escalate. Water, especially fresh water, defines life on this planet. Demands for it are increasing in all use areas, especially agriculture, where it is used for both production and food processing. Crop plants require copious amounts of water. Estimates are that 4,000–6,000 gallons of water are needed to produce a bushel of corn. That translates into 800,000–1.2 million gallons of water to produce 200 bu/A corn.

Historically, irrigation was the largest user of water. In the United States in 2005, however, cooling for thermoelectric power generation was the largest user of water (49%), accounting for about half of the 410 billion gallons per day withdrawn, 92% of that used on a once-through[4] basis (Barber, *Summary of Estimated Water Use*). Irrigation was the second largest use, accounting for 39% of the total, and public water use, industrial uses, aquaculture, and livestock uses constituted the balance. Water not withdrawn from rivers and streams provides important environmental services, including the required protection of endangered aquatic species.

Although the magnitude and causes of adjustments in climate and impact on agriculture are still being debated, most people accept that changes are occurring. Most recognize there will be a mix of environmental changes—some good and some bad, but most agree water will have one of the greatest impacts. For example, some regions will be warmer and some colder, some wetter while others will be drier, and new crops will be grown in regions where they have not been previously grown. The challenge for agricultural research is to develop crop cultivars and crop- and livestock-production systems that enable successful agricultural production regardless of the specific nature of the climatic changes. If agricultural scientists accept this premise, then research programs can minimize the effect of climate change.

Several approaches are possible. For example, breeding plants with greater drought tolerance is a long-sought-after goal. Water is an increasingly important factor in agricultural productivity in many regions of the world. Estimates are that 40% of corn crop losses are due to lack of water. Water will become a more serious factor affecting productivity with the further progression of global climate change. Considerable research is under way that holds promise for conferring remarkable levels of drought tolerance on corn. Similar gains are being sought in animal agriculture. Developing livestock that can tolerate and indeed thrive under adverse conditions would strengthen animal agriculture.

Example 7: Developing Crop Plants That Are More Effective and Efficient in Using Plant Nutrients

Developing crop cultivars that are more efficient in using plant nutrients is another important challenge. Nitrogen is described as one of the grand challenges, but other plant nutrients are also of concern. The availability of nitrogen fertilizer is linked primarily to fossil energy—natural gas. Consequently, improving the efficiency of plants' use of nitrogen would free up natural gas for other uses. Phosphorus derived from phosphate rock is a limiting mineral resource in crop production. The United States extracted 31 million metric tons (mmt) of phosphate rock in 2008 from a reserve base of 3,400 mmt, or a 110-year supply at the 2008 rate of use (Buchanan, Herdt, and

Tweeten, *Agricultural Productivity Strategies*). This supply is not a comfortable margin for an element so basic to crop production, and the United States eventually will become a net importer of phosphate.

World phosphate production totaled 167 mmt in 2008 from a reserve base totaling 47,000 mmt, or a 281-year supply at the 2008 production rate. Nearly half of the world's phosphate rock reserves are in Morocco and Western Sahara. The research challenge is to develop crop cultivars that more efficiently use such critical factors of production. Fortunately, we have huge reserves of potassium, so this nutrient will not be limiting in the foreseeable future.

Example 8: Improving Pest Resistance in Plants and Livestock

Modern agriculture that involves large acreages of monocultures or large numbers of animals confined in proximity creates an ideal environment for pests. The standard definition of a pest is "anything that causes trouble, annoyance, discomfort, etc.; specifically, any destructive insect or other small animal; vermin" (Oerke and Dehne, "Safeguarding Production"). In agriculture this definition may be sharpened just a bit to include anything that detracts from, diminishes, or destroys a crop or animal cultivated by people.

Crop and livestock losses because of pests represent one of the most serious impediments to agricultural productivity. Crop losses are caused by a range of organisms, including weeds, insects, and other animal pests, pathogens, viruses, and so on. Livestock losses are caused by a range of agents from viruses to insects and other organisms. Hardly any crop or livestock enterprise is free from pests. Losses caused by pests vary substantially by crop or livestock enterprise, region, and production practice. Actual crop losses caused by various agents are estimated at 26–30% for sugar beets, barley, soybeans, wheat, and cotton (ibid.). Losses for corn, potatoes, and rice are estimated at 35%, 36%, and 40%, respectively. Developing means of avoiding or minimizing losses caused by plant and animal pests holds great promise for both crops and food animals. Consequently, this represents one of the important challenges in agricultural research.

Example 9: Elimination of Diseases of Animals

Although improving resistance to pests by food animals addresses one of the important challenges in agricultural research, the elimination, or at least mitigation, of the effect of animal diseases would have a dramatic impact on agricultural productivity. Losses in productivity of livestock vary widely by country. As with crops, numerous agents, including pathogens, affect animals. For example, rinderpest, also called cattle plague, is an infectious viral disease of cattle and some other animals. An outbreak in the 1890s is thought to have killed 80–90% of the cattle in southern Africa. As late as the 1980s an outbreak in Africa caused a half-billion-dollar loss. Today the world is free of rinderpest. The last confirmed case was in Kenya in 2001 (Normile, "Driven to Extinction"). The last surveillance operation took place in 2009.

This example illustrates what is possible. Obviously, much dedicated and highly focused research is required to achieve such an outcome, but our future is predicated on such developments. The commitment to research on human disease is impressive. Both public research institutions, such as the National Institutes of Health (NIH), and private organizations, such as the Bill and Melinda Gates Foundation, are highly committed to improving the lives of people through research on a wide range of diseases.

A similar commitment to solving animal diseases should be made for two important reasons. First, food animals constitute a significant portion of our diet. Second, several human diseases are linked to some diseases of animals.

Example 10: Improved Nutrition of All Species of Food Animals (Land and Water) Constitutes an Important Aspect of Food Security for Humankind

Research that leads to more efficient use of feed for food animals holds great potential. Much research has focused on more effective use of nonhuman food products such as animal feed from oilseed by-products, ethanol production, rendering, and so on. Animal diets that minimize the inclusion of grains or other feed that could be consumed by humans should be a goal of research.

Past agricultural research has been highly successful, particularly with some species such as poultry. Today it is expected that a marketable broiler chicken weighing 6 pounds can be grown in 45 days with a feed-conversion ratio of 1.85 or less. Other research has focused on more extensive use of grass in beef production. In fact, some 100% grass-fed beef is now available in certain upscale grocery outlets.

Example 11: Developing and, Where Possible, Incorporating the Hybridization of Crops

Hybrid vigor has enabled some crops to achieve a higher level of productivity as evidenced by hybrid varieties of corn. Unfortunately, the development of hybrids in rice, wheat, cotton, and so on is much more difficult. Hybrid seed is expensive because of the requirements for planting different lines to produce the hybrid seed itself. However, the further development of hybrids in other crops is still desirable since improvements brought about by hybridization provide a highly efficient means of crop improvement.

Example 12: Incorporating the Process of Apomixis into Crop Plants

The requirement of annual hybrid-seed production can be circumvented by the process of apomixis, which is the production of seed without fertilization by the male gamete in pollen grains, resulting in progenies identical to the seed-bearing hybrid plant. Hybrid vigor has enabled some crops to achieve a very high level of productivity as evidenced by hybrid varieties of corn. Without the annual reproduction of first-generation hybrid seed, productivity in subsequent generations would continually decline because of loss of heterosis. Apomixis would enable hybrid crop plants to maintain hybrid vigor at no additional cost (i.e., the annual hybrid-seed-production field is unnecessary if the hybrid plant is apomictic.)

Example 13: Improving Plant Response to Elevated Levels of CO_2

The early to mid-eighteenth century and the beginning of the Industrial Revolution witnessed a dramatic increase in greenhouse gases, particularly CO_2. This increase was due primarily to burning coal and, later, petroleum. Prior to the industrial revolution CO_2 in the atmosphere was 280 ppm. Today it measures about 390 ppm in many regions. It is expected that CO_2 will continue to increase in the foreseeable future. What is already understood about photosynthesis suggests a variety of refinements that could take advantage of this expected rise. Research aimed at understanding the determinants and improving the responsiveness of crops to elevated levels of CO_2 would be valuable.

Example 14: Improving Energy Efficiency of Plants

One of the primary means of harnessing the sun's energy is green-plant photosynthesis. Unfortunately, plants are quite inefficient in capturing energy from the sun. Calculations show that less than 3% of sunlight absorbed by the leaf is converted to chemical energy; frequently, it is less than 1%. This number is in contrast to photovoltaic cells (solar panels), which routinely capture 10–15% of light. With new technology in solar panels, efficiency may become even greater, up to 20%. Most crops capture light in the range of 400–700 nanometers (nm) in wavelengths. On the other hand, throughout nature certain organisms capture light from 400 to 900 nm. The light-harvesting mechanism (consisting of grana) of crops could be improved through genetic engineering. It is conceivable that genes could be transferred to improve energy efficiency either by enhancing the light-harvesting capabilities or by expanding the wavelength of the light being captured.

Plants have evolved a sophisticated photo-protective mechanism that kicks in when they receive more light than they are able to use in photosynthesis. While the precise nature of the photo-protection process is not clear, excessive protection could decrease photosynthetic efficiency. Refining and developing a better understanding of the photo-protective process could lead to enhanced photosynthesis. Clearly, improving the energy efficiency of crop plants could lead to strengthening agricultural productivity.

Example 15: Developing Commodities with Increased Health Benefits

Another means of enhancing agricultural productivity is the development of nutritionally enhanced food commodities, perhaps including those that have unique properties such as increased health benefits. Traits such as specific amino-acid profiles, unique proteins, particular antioxidants, vitamins, or minerals would enhance the value of a commodity to the consuming public. An exciting prospect is designing plants that can synthesize medicinal products. This would truly be an important innovation.

Example 16: Developing Plants That Can Tolerate or Avoid Stress

Given the potential for global climate change and all of the unknown impacts on the environment, improving plant and animal tolerance to adverse growing conditions becomes an important consideration. For example, breeding plants that are drought or flood tolerant, plants that thrive in colder or hotter climates, or plants that tolerate salts are possibilities.

Soil salinity is one of the major abiotic stressors affecting agricultural productivity in many parts of the world. The problem is exacerbated when irrigating with water that has high salt concentrations. Two main approaches to improving crop salt tolerance are (a) developing more salt-tolerant cultivars using naturally occurring genetic variation either through direct selection in stressful environments or by mapping quantitative trait loci and (b) subsequent marker-assisted selection. Another approach is to generate transgenic plants to introduce novel genes or to alter the expression levels of existing genes.

Development of plants that have more tolerance to drought, higher temperatures, flooding, and so on is challenging research. However, when one realizes that many such traits are already embedded in a number of plant species throughout the plant kingdom, the challenge seems more achievable. The goal is to develop plants that not only survive under such adverse conditions but actually thrive under them. For example, the Sub1 gene in rice allows rice to recover nearly completely if submerged for no longer than two weeks, a tremendous breakthrough in accommodating the plant to the environment.

Though we still do not have a clear picture of precisely how global climate change will affect our environment, we need to focus on growing plants and livestock under a wide range of environmental conditions.

Example 17: Delaying Plant Senescence

Agricultural productivity is determined to a great extent by photosynthetic efficiency, the amount of light captured and used, and the duration of the photosynthetic period. Each of these areas has the potential to strengthen agricultural productivity. By delaying senescence, that is, by keeping plants green and continuing to photosynthesize, a greater amount of fixed carbon would result and thereby enhance productivity. Much research centers around various aspects of senescence in humans. If this work results in delaying senescence in humans, many social ramifications will occur. On the other hand, delaying senescence in plants is a win-win situation.

Example 18: Effectively and Efficiently Preventing or Capturing All Animal Waste

Developing means of capturing methane gases or utilizing animal manures would strengthen agricultural productivity. The research challenge is to develop innovative ways to capture the value of animal gases and manures. The idea of using animal manure is not a new idea. Indeed, animal manure has long been recognized as a nutrient-rich amendment for crops. However, in recent years some improvement has occurred in the capture of the energy in such material. More definitive research could lead to even more efficient means of animal waste use.

Example 19: Using the Power of Genomics and Biotechnology to Improve the Quality of Food Animals

The power of genomics, proteomics, and other aspects of biotechnology offer great potential for changing the dynamics of animal agriculture. Indeed, so far we have only scratched the surface as to the overall potential. While classical breeding of food animals has made significant improvement in their quality, great potential remains for still further improvement. Such research could address health issues as well as the resulting food products.

Example 20: Seeking New Innovations That Offer Possibilities

We must seek creative and innovative approaches to agricultural productivity in the future. One of the possibilities for powering the next green revolution is "farming" the world's oceans. Since much photosynthesis is ongoing in the tremendous expanse of the oceans, developing farming practices for ocean crops may well provide a quantum increase in agricultural output. For instance, bioengineered algae that convert waste from coal-fired plants into biofuel may be one such possibility.

The potential for an ocean-based grain crop is exciting to contemplate, and the immediate reaction that "it's never been done" is not unexpected. It makes as much sense to seek enhanced food production in the world's oceans as in outer space. Of course, some type of food production in space will become needed when interstellar flights spanning many years become a reality.

When considering farming the world's oceans, one recognizes the countless problems it will involve—from developing plants that produce some type of seed or other form of useful, concentrated food product in the oceanic environment, to devising an aquatic combine! But solving such problems is the nature of research.

Reclaiming land areas that have various problems such as deep sands, polluted soils, and coastal areas for crop production may help improve the environment and enhance agricultural productivity. Other innovative approaches also deserve consideration. Our effective use of valuable land resources could be vastly improved. For example, highway and power-line rights of way require many acres of land, some of which is highly productive and suitable for producing food or biomass for fuel.

Opportunities outlined in this example require that agricultural scientists (and administrators) get out of their comfort zone and design research projects that might lead to breakthroughs in food production rather than just kicking the can down the road.

Example 21: Developing Innovative Methods for Water Harvest

Certain areas of the United States receive considerable rainfall each year. More than 50 inches per year is an adequate amount for the production of most crops—if properly distributed during crop growth and development. The obvious solution is to capture or "harvest" water when it is not needed and to store it for later use. In contrast to the innovative approaches that have been developed and are used in drier regions, little definitive research has focused on water harvest in other regions other than building small farm ponds or large reservoirs. Innovative approaches should consider both small-scale (on farms) harvesting approaches as well as large-scale (watershed) approaches. In addition to developing means of harvesting, new storage options are needed. This is hardly a novel idea. In fact, in many countries where water is scarce, this has long been an ongoing process. Consequently, areas where water receipt during the growing season is the primary problem can use approaches that concentrate on water storage.

Many private homes that are being built today include a cistern to catch rainwater, which can be used for irrigating the lawn, shrubbery, or perhaps a home garden. This is also not a new idea; cisterns were common 75 years ago—before electricity reached rural areas. Such water then was primarily for livestock or emergency use to put out fires.

Example 22: Improving Storage of Food

Some estimates are that as much as 30% of the available human food is not consumed. A report from the Natural Resources Defense Council (NRDC) issued in August 2012 mentions that "up to 40% of US food from the farm ends up in landfills." The cost of this uneaten food is $165 billion each year, and it is the largest component of municipal waste (Gunders, "Wasted"). There are numerous reasons for this simple fact. In some countries lack of refrigeration may lead to spoilage. Lack of protection from insects, mice, and other animals that subsist on human food causes tremendous loss. Another quite avoidable loss is due to consumers who take "sell-by dates" all too seriously. This, along with the excessive caution exercised by food processors with regard to the shelf life of their product, causes much good food to be tossed out on the exact sell-by date. In view of liability concerns, grocery stores usually adhere to sell-by dates. Reducing such loss would go a long way to meeting the food needs of the planet by 2050. Rather than using an arbitrary sell-by date to dispose of food, some effort should be made to develop a means of determining when a food product is no longer safe to consume.

Each of these grand challenges or ideas presents researchable problems of a Herculean nature. Solving any of these would be a tremendous breakthrough. Indeed, they are the kind of problems on which much of our agricultural research effort should be focusing. These are the types of problems we must solve in order to ensure the future success of agriculture and the well-being of the people on this planet. The challenges are real and daunting—but not impossible with dedicated and focused research.

In a future world, where demands for energy are more closely aligned with availability, agriculture must seek higher levels of efficiency. Successfully addressing many of these challenges would contribute to achieving a greater and more sustainable efficiency of agricultural production and get us on track to meet the future food and energy needs of the planet.

10
Federal Funding for Research: Agricultural versus Other Research Programs

You must never believe all these things which the scientists say because they always want more than they can get— they are never satisfied. [Responding to a complaint of inadequate support for research]

—NIKITA SERGEYEVICH KHRUSHCHEV

Emergence of Research Funding for Agriculture

The human endeavor we now call research encompasses studies that seek to understand natural physical and biological phenomena and efforts to use that knowledge to improve the lot of ourselves and others. From the earliest times, such studies (we sometimes refer to these as basic research) were driven by curiosity. In general its funding was a personal and private matter, left to wealthy individuals or to churches, much as art was funded. This was understandable. Though the results of basic research added to our body of knowledge and laid the groundwork necessary for future advancements, they were, and often still are, intangible. Payback or return on investment was often indiscernible or unrealized by the funder. Nonetheless, the quest to understand the world has existed in many cultures since at least recorded time.

However, humans also seek to improve their circumstances, provide better comfort and security, or generate wealth, whether that was grain in storage, land under their control, or gold in their vaults. To achieve these things, they have systematically applied novel approaches using what they had learned through experience, study, and deliberation. The earliest documented forays into research were made in Egypt, Mesopotamia, and China in the ninth and tenth centuries. This was followed by the Europeans in the fifteenth and sixteenth centuries. However, these efforts did not involve agriculture. Scientific agricultural investigation that effectively began in the mid-nineteenth century ushered in a new age. Nevertheless, the tremendous contribution made by agricultural research is often overlooked. Agricultural research provides a clear illustration of the power of focused studies in many endeavors. As with basic research and art, applied agricultural research was often funded by individuals and institutions that were not (primarily) governmental. Early agricultural research was also often supported by a benefactor. For example, as explained in chapter 3, the Rothamsted Experimental Station was initially financed wholly by Lawes and only later by public monies through the Agricultural Research Council of the United Kingdom (*Rothamsted Experimental Station*).

Gregor Mendel was an Austrian Augustinian monk who was born in 1822 and died in 1884. His affectionate epithet, "father of modern genetics," is well deserved because of his research with peas. At the time of his work, however, he was little recognized. But by the early twentieth century, the importance of his work was realized. Mendel's research with peas enabled him to make two basic contributions to genetics: (1) the law of segregation and (2) the law of assortment. The latter became known as Mendel's law of inheritance.

Because Mendel was a monk, his research was supported by the church. In 1868, at the height of his scientific career, he was elevated to the position of abbot. Unfortunately, this essentially brought an end to his research career. Mendel might have been the first, but he certainly was not the last good scientist who was lost to administration. Funding by various institutions and organizations was a common means of supporting agricultural research through the early and mid-nineteenth century.

Church funding of research in Europe is an interesting case that parallels later governmental funding. Though organized for the propagation of religious thought and theology, churches were de facto authorities throughout Europe during and shortly after the Middle Ages. Though their fundamental mission was to provide insight into and avenues to God, they recognized that they could make little headway among a population that was starving, ill, and unable to read or reason. They provided food to the poor but, more so, saw that they needed to teach them how to provide food for themselves. They established care for those who were sick and built early hospitals but saw that they also needed to understand the human body and how it was affected by illness. They also established many of the early educational institutions that not only taught people how to read and think but also unleashed their curiosity and added structure to their endeavors in the basic sciences. This role of support for education and research, which seeks to improve people's lives, became the basis for public education and the public funding of research, which proved successful in changing the United States from a collection of former colonies primarily engaged in farming into a world power.

In another parallel that persists today, early scientists found that discoveries or research results that went against the established ideas and opinions of their patrons and funders were met with more than a little resistance. Though Galileo's ideas caused him trouble in his own lifetime, the church eventually found that they really posed little threat to the church's authority and beliefs. Not so with Charles Darwin. His research findings still create harsh criticism from some groups. Secular institutions and governments have often found themselves acting as apologists trying to appease the church's views while funding research that is still built upon Darwin's ideas.

In most cases, agricultural research seldom stepped on religious fundamentals. Whether institutions and scientists were church supported, privately supported, or supported by emerging governments of the nineteenth century, applied research was viewed as good for everyone. The realization that a community of people can improve their livelihood through new practices and discoveries is infectious. What was learned in one place was adopted elsewhere, and others built upon the most recent ideas.

The highest economic or material interests of a country, the increased and more profitable production of food for man and animals . . . are most closely linked with the advancement and diffusion of the natural sciences, especially of chemistry.
—JUSTUS VON LIEBIG

Following the development of agricultural research in England at places like Rothamsted, attention generally shifted to Europe, particularly Germany. This occurred soon after Justus Liebig's landmark publication, *Organic Chemistry in Its Application to Agriculture and Physiology*. German universities established a strong international reputation as centers of learning in agricultural science. Under Bismarck, Germany funded a set of agricultural experiment stations. Other countries, especially the United States, sought to emulate the German and English models when it created its own system of state agricultural experiment stations.

States versus Federal Roles in Research: Funding versus Research Direction

Since the United States was still a federation of states in the 1800s, it was mostly individual state governments that sought to bring the emerging European model of education and research to agriculture. Even in the Northeast, most of the citizens lived in rural communities and were engaged in agriculture or agricultural support. It is no wonder that the states realized that practical discoveries in agriculture would enhance their citizens' lives and improve the state's economy.

The first step in centralizing research and education in support of agriculture came shortly after the Civil War. In 1862, Congress passed the Morrill Act, creating what was known as the land-grant system. In essence each state was called upon to designate an existing educational institution or create a new one that would provide an educational opportunity to learn about the latest techniques of farming and engineering. At the time, much of the engineering effort was applied to agricultural practices and equipment or equipment used in the handling and processing of agricultural commodities. This action by a newly emerging federal government would invest in the people of each state and help restore the agricultural and economic base of those states ravaged by the Civil War.

Emergence of the nation's land-grant universities (see chapter 2) provided a great stimulus to research. As these programs progressed, it became obvious that teachers needed more information. Such efforts followed many paths in different states. About the only thing these research efforts had in common was their meager funding bases. Almost everyone desired the results of the research, but few were willing to provide financial backing. Since quality fertilizer material was one of the most common concerns among farmers, some experiment stations evolved as a means of analyzing materials that were offered for sale as fertilizers. Fees charged for this service provided a small (often very small) funding stream for the research institution.

This piecemeal, noncoordinated approach to funding was unable to sustain the kind of research that was necessary to address many of the emerging problems in agriculture, which was expanding across the rapidly growing country. The first coordinated federal effort came with the passage of the Hatch Act of 1887, which created the state agricultural experiment station system (see chapter 3). It provided modest funding for each state in the amount of $15,000 per annum. From this inauspicious beginning would arise one of the most impressive agricultural support systems ever conceived.

What seems so simple and straightforward can become exceedingly complex when dealing with several sovereign states and a federal government that attempts to seek unity among all states. First, it was the intent that each experiment station would be affiliated with a land-grant university in each state. For different reasons, not all states wanted the experiment station associated with the land-grant university. Second, an issue that still rankles some, particularly in the federal bureaucracy, is the simple fact that, even though the Hatch appropriation is federal money, the respective states set their own research priorities. To those of us who have spent our life in the university system, this is logical and readily accepted. It seems only reasonable that those nearest to the problems can most effectively direct the research needed to solve them. However, individuals who come into the system from other endeavors often have great difficulty appreciating this relationship.

Even though agricultural research occurring in the states receives federal funds, the execution and results of such research are not under federal jurisdiction. Once money has been awarded to the states, all of the research decisions and results remain in their purview. Before the first payment could be made to the states under the 1887

Hatch Act, the US comptroller of the treasury ruled that the Hatch legislation provided no financial appropriations (Kerr, *Legacy*). To address this problem Congress passed special legislation to provide money to implement the Hatch Act. Passage and funding of the 1887 Hatch Act was a beginning, although a feeble one. Soon everyone, especially Congress, recognized that, to be effective, more resources were needed for financial support of the experiment stations. Over the years several congressional legislative acts were passed that continued to improve the resource base for agricultural research. These legislative acts are more fully described in chapter 3.

At the time of the passage of the 1862 Morrill Act, which created the land-grant system, the United States was in a unique situation. At the conclusion of the US Civil War many things were decided, but not everything was changed. The Emancipation Proclamation made all people free but did not guarantee all people equality. The land-grant concept was a great idea. Unfortunately, in the Confederate states, the African Americans had their freedom but certainly were still disenfranchised in many ways, particularly in higher education. To remedy this situation, Congress enacted legislation often referred to as the Second Morrill Act, which was approved on August 30, 1890 (Tegene et al., *Investing in People*). Institutions made possible by this legislation often are referred to simply as 1890 institutions or historically black land-grant colleges. This legislation provided for most of the provisions found in the 1862 legislation to the people of color in the states of the Old Confederacy. Making the 1890 institutions full and equal partners with the 1862 institutions is still a work in progress. Nearly a century later, Congress provided support for the "1994 land-grant colleges" for Native Americans. Incorporating the 1994 land-grant institution into the land-grant system is also under way.

From the beginning, it was recognized that agriculture was important from both national and state perspectives. Consequently, it was envisioned that both the federal government and the state governments would provide support. The required matching of federal dollars with state dollars ensured a joint federal-state partnership. While the requirement for at least a one-to-one match has long been required for the 1862 experiment stations, the requirements for at least a one-to-one match in the 1890 experiment stations had a more recent vintage, beginning with a 30, 45, and 50% match in fiscal years 2000, 2001, and 2002, respectively.

Additionally, among both the 1890 and the 1862 land-grant institutions, a great discrepancy among states has existed in the magnitude of the matching appropriations. Some states have struggled to match the federal appropriation, whereas other state legislatures have appropriated monies that exceeded the requirement by a considerable amount.

From the earliest days of this country, power within the Union has been greatly debated—who has it and how it should be divided between the states and the federal government. Not surprisingly, some of these early disputes were about money. Some of the founders who were Federalists included George Washington, John Adams, John Hancock, Alexander Hamilton, and James Madison. They believed in greater power for the federal government, whereas others, considered as antifederalists, including Sam Adams, Richard Henry Lee, George Mason, and Patrick Henry, were committed to limiting the power of the federal government. Thomas Jefferson, another founder, was away during some of these deliberations and was not openly committed to either point of view but probably sided with the Federalists after a Bill of Rights was included and the Constitution was ratified. These issues continue to be debated today. One thing that is still a factor is the federal government's tendency to take the lead and fund activities focused on matters that concern the common good of citizens in many states.

Federal Support of Newer Research Initiatives vis-à-vis Agriculture

While agricultural research and education continued to struggle with the issue of state versus federal control, the success of the enterprise was never in doubt. The research paid dividends that could be measured not only in the lives of farmers but also in the health and well-being of the populace. Each improvement in efficiency enabled more people to be freed from the land to tackle other endeavors that added to the wealth and strength of the US economy. That very success led Congress to understand that federal investments in basic and applied research could stimulate other sectors of the economy.

Over the next few decades this idea proved its soundness as research in many areas began receiving federal support. During the twentieth century, federal backing emerged and has grown in several federal agencies, including the Departments of Defense and Energy, the National Institutes of Health, the National Science Foundation, and the National Institute of Standards and Technology (NIST). More recently, funding for research was initiated for the Environmental Protection Agency and the Department of Homeland Security.

It is instructive to review some of these major federal funding initiatives before we consider agriculture.

National Institutes of Health

One of the most impressive developments in the funding of research by the US government is the National Institutes of Health (NIH), which began in 1938 with an initial appropriation of $464,000; however, in just 5 years the appropriation almost tripled and within a decade had grown to more than $24 million. In 1965, funding for NIH reached almost $1 billion per year (table 1). By 1975, appropriations had doubled to more than $2 billion.

Between 1977 and 1987, appropriations for NIH almost tripled, to exceed $6 billion per annum. In the decade 1986–1996, federal appropriations for NIH increased to $11.9 billion, again almost doubling. But the best is yet to come. The decade 1996–2006 saw the budget for NIH climb to more than $28 billion, with growth continuing even considering the poor status of the US economy.

In 2010, the federal appropriation for the NIH stood at $31,238,000,000. With the US population at slightly above 300 million, this means that each person in the United States was receiving the equivalent of $100 per annum of federal dollars for medical research and related support from the NIH. On the other hand, federal support for agricultural research provides only about $10 in such support for each person in the nation.

Table 1.Appropriations for Selected Federal Agencies in 5-Year Increments for the Period 1955–2010 (Dollars in Thousands)

Year	NIH[a]	NSF[b]	NIFA[c, d]	ARS[e]	FS[f]	ERS[g]
1955	80,151	12,490	19,595	36,967	7,296	–
1960	399,380	158,600	32,131	68,266	14,545	–
1965	959,159	415,970	49,999	128,890	35,510	–
1970	1,061,007	462,490	62,640	159,220	44,424	18,420
1975	2,092,897	693,130	101,749	224,323	75,402	26,333
1980	3,428,935	975,130	186,031	372,970	111,531	42,503
1985	5,149,459	1,507,070	304,316	492,022	134,483	48,598
1990	7,576,352	2,026,060	385,586	585,923	159,176	52,688
1995	11,299,522	3,270,270	493,040	713,866	211,180	63,661
2000	17,840,587	3,923,360	487,462	830,384	217,694	67,878
2005	28,495,100	5,480,770	669,579	1,102,000	276,929	75,796
2010	31,238,000	6,872,510	808,390	1,179,639	306,637	87,225

[a] National Institutes of Health, Office of Budget, History of Congressional Appropriations.
[b] Retrieved March 18, 2013, from http://dellweb.bfa.nsf.gov/NSFHist.htm.
[c] National Institute of Food and Agriculture, Research and Education Activities, Appropriation History, retrieved May 17, 2015, from http://192.73.224.129/about/offices/budget/10yr_res_ed.pdf.
[d] These data include research and education appropriations.
[e] ARS Historical Budget Files.
[f] Forest Service Historical Budget Files.
[g] ERS Historical Budget Files.

National Science Foundation

The National Science Foundation (NSF) supports fundamental research and education in all fields of science and engineering with the exception of medical science. Thus, by definition, the NSF bears some responsibility for agricultural research. Within NSF are six research directorates and one education directorate, but none is devoted specifically to agriculture. However, one directorate is devoted to the biological sciences. This directorate, as well as others, serves as a potential source of assistance for agricultural research. However, bear in mind that many biological science research challenges and opportunities exist in addition to agriculture. With the minor role that agricultural research plays in the greater research endeavor in the United States today, it is important to recall the leading role agriculture played in fostering research in many areas.

In the twentieth century, World War II contributed to the phenomenal growth of research. Certainly, the success of research in many areas was a factor in the final outcome of this conflict. Much war-related research led to numerous innovations based on science and technology. In fact, the Manhattan Project led to development of the nuclear bomb, which brought an abrupt end to World War II.

It was clear to many national leaders that science and technology were major contributors to the successful outcome of the war. Consequently, they were interested in maintaining these advantages. After much wrangling in Congress, the National Science Foundation was created. The obvious contributions of research to the war effort provided a great stimulus for experiments in many fields, particularly those that supported national defense. An example of agriculture's contribution was the role of the USDA's Agricultural Research Service in developing a means of producing large quantities of penicillin. Alexander Fleming, a Scottish bacteriologist, had discovered the great therapeutic value of penicillin but was unable to produce the antibiotic in large quantities. Scientists at the USDA-ARS Northern Laboratory in Peoria, Illinois, were able to make numerous improvements that led to the mass production of penicillin. Obviously, this was a great contribution to the war effort as well as a benefit to all humankind.

The NSF came into existence just as the United States was becoming engaged in another conflict—the Korean War. Even after approval of the concept, the first NSF budget was a modest $151,000 in 1951. Increasing steadily, the NSF budget exceeded $100 million before the end of the decade (table 1). It would reach almost a half billion dollars by the end of the second decade after it was established and a billion dollars by the end of the third decade. By the end of the fourth decade the NSF appropriations were more than $2 billion and almost $4 billion by the end of the fifth decade from its founding. By 2010, the NSF budget appropriations had grown to almost $7 billion, with modest increases in 2011. The NSF received $2.4 billion (one-time money) from the American Recovery and Reinvestment Act (the "stimulus bill") in 2009. Most of the additional research funding provided by the stimulus was for capital improvements at research institutions.

US Department of Agriculture

In addressing federal appropriations for agricultural research from a historical perspective, one immediately encounters a challenge. First, unlike NSF and NIH the research component of agriculture is apportioned among several agencies in the USDA, not all of which are focused on research. Second, over the years various agricultural research components of the USDA have undergone reorganization and restructuring numerous times. An excellent review and summary of federal agricultural research is found in *An Assessment of the United States Food and Agricultural Research System,* prepared by the Office of Technology Assessment at the request of Congress.

Before we can examine the levels of federal research funding in agriculture, we need to review the agricultural research agencies. The focus here is on the more recent history of the organizational structure and appropriations for the current system rather than on a full 150-year historical assessment of appropriations for agricultural research. Consequently, consideration of organization and structure will begin with the reorganization of 1953, which brought about the abolishment of the long-standing bureaus and the Office of Experiment Stations.

The reorganization of 1953 brought about one notable miscue: placing the cooperative state research programs under the federal Agricultural Research Service. This was rectified in 1962 with the establishment of the Cooperative State Research Service (CSRS). In view of the many diverse interests in agricultural research, it is not surprising that wide concern continued with regard to the organization and structure of the research enterprise.

During the 1960s and 1970s various groups offered advice to improve the system. Among those was the Life Sciences Panel of the President's Science Advisory Committee and the National Academy of Science Committee on Research Advisory to USDA—(later known as the Pound committee). While these groups were all well meaning and offered excellent advice along with some criticism, it was left to the Food and Agricultural Act of 1977 (the Farm Bill) to provide the next organization and structure for "improvement" of agricultural research. The 1977 Farm Bill provided for a joint council consisting of the Food and Agricultural Sciences and the National Agricultural Research and Extension Users Advisory Board. In addition to these "improvements," the Science and Education Administration (SEA) was established within the USDA and was given the authority to carry out all agricultural research, extension, and education efforts.

The next Farm Bill, passed in 1981, provided for an assistant secretary to oversee all research and education components of USDA. Such a position had been a long-sought-after goal of the state agricultural experiment station directors and associated organizations. Most of the directors surmised that such a position within the USDA would provide for greater recognition of and support for agricultural research. They preferred a single point of contact in the federal government for all agricultural research funding. However, even before the new assistant secretary was appointed, the "shotgun marriage" of the USDA agencies into the Science and Education Administration was dissolved (Kerr, *Legacy*). The Agricultural Research Service (ARS), the National Agricultural Library, the Cooperative Extension Service, and the Cooperative State Research Service once again became stand-alone agencies. Also, in 1981, the Economics and Statistics Service was reorganized into two separate agencies—the Economic Research Service (ERS) and the National Agricultural Statistics Service (NASS). The CSRS later merged with the Federal Extension Service to become the Cooperative State Research, Education, and Extension Service (CSREES). Later, the 2008 Farm Bill provided for the creation of the National Institute of Food and Agriculture (NIFA), which essentially replaced CSREES (Food, Conservation, and Energy Act of 2008).

This overview shows the multitude of changes in and reorganizations of the agricultural research effort. For these and other reasons, it is difficult to make budget comparisons. Complicating the matter even further is that some of the appropriations information includes research, along with other endeavors such as education and extension. The appropriation for NIFA is probably the most complex of all of the agricultural research agencies.

Agricultural research was among the first investigative endeavors supported by the federal government. In 1940, the federal appropriation for research in the state agricultural experiment station (SAES) systems was slightly more than $7 million.[1] Modest growth in the appropriations occurred for the next 15 years. Appropriations in 1955 brought about a doubling of support for agricultural research (table 1). It only took 7 years for the appropriation to double again. Doubling again occurred in nine, seven, and eight years; after a few lean years in the late 1980s, growth resumed and broke the $500 million mark in 2001. It took 20 years, 1990 to 2010, for the appropriation for the SAES to double once again.

Allocation of resources to the state agricultural experiment stations according to the Hatch formula varies greatly. One is immediately struck by the modest allocations of Hatch funds for agricultural megastates such as California and Florida when compared to less agriculturally productive states. Many changes have taken place in agriculture and other parameters in the Hatch formula, while unfortunately the formula has not changed appreciably.

ARS

The initial appropriation for the Agricultural Research Service in 1953 was $32,922,000 and was doubled in 7 years (table 1). With modest yet constant growth, appropriations doubled again in 6 years, and again in 10 and 13 years. However, it would be 17 years before the ARS appropriations would double once more. The appropriations for ARS stood at $1,179,639,880 in 2010. Modest budget reductions have occurred in recent years.

ERS

The economic, research, and statistical collection agencies had been merged into the Economics, Statistics, and Cooperatives Service (ESCS) in an earlier (1977) reorganization. The 1981 reorganization of ESCS separated its economic research and analysis function from its statistical collection and analysis functions. This enabled the ERS to be a stand-alone agency. At the time of the separation, the appropriation for the ERS was $44,698,000. By 1990, the appropriation had grown to more than $50,000,000 and reached $67,878,000 in 2000 (table 1). The ERS budget exceeded $87,000,000 in 2010.

During this period, when federal appropriations were generally increasing, both personnel and appropriations for agriculture were undergoing dramatic reductions in real dollars. While ERS staff years dropped from about 1,000 in 1981 to about 400 in 2011, appropriations grew from about $44 million in 1981 to slightly more than $83 million in 2011.

Forestry

Forestry is generally thought of, particularly by those in traditional agriculture, as a major component of agriculture. Some, mostly those who consider themselves foresters, see forestry as a separate entity altogether or as a major component of natural resources. Regardless of how one perceives forestry, it is important from many standpoints. In view of the dichotomy of perspective, appropriations for forestry research reside in both traditional agriculture and the natural resource areas. Primary funding for forestry research in the US Forest Service is in the natural resources and environment mission area of the USDA. Funding for forestry research in the state agricultural experiment stations is provided by the McIntire-Stennis legislation. While these funds are administered through the agricultural experiment stations, they are often delegated to a school of forestry or a separate department of forestry. Sometimes these funds are administered by a completely separate institution—not even a land-grant institution.

Appropriations for forestry research in the US Forest Service for the period 1995–2010 are shown in table 1. Earlier appropriations for forestry research beginning in 1881 were extremely modest and remained so for the remainder of the nineteenth century. Research support for forestry grew steadily during the early part of the twentieth century and reached more than $1 million in 1929. Some growth occurred for the next few years but was only slightly above $1 million in 1934. Appropriations for forestry research increased during the 1930s, '40s, and '50s. Beginning in the 1960s, funding for forestry research continued to increase, reaching more than $200 million by the year 2000 and over $300 million by 2010 (table 1).

The foregoing discussion illustrates several important points. The US government's financial support for agricultural research, particularly in recent years, is critical for the future of agriculture. It is dynamic in embracing new areas and can move quickly when a new need arises. As a citizen of this country and as a scientist I applaud this commitment to the development of information, knowledge, and technology pertaining to agriculture. I can make a sound case for support of the total portfolio of federally supported research. The research agencies NSF, NIH, the Department of Energy's Office of Science, and others address many areas that affect our lives in a multitude of ways. A similar argument could be made for their counterparts in the EPA, DOD, National Institute of Standards and Technology, and NASA budgets devoted to research.

Another inescapable fact is that generally substantial positive growth in appropriations has occurred in most areas of research. The exception is agriculture. One wonders why agricultural research appropriations have not kept pace with other areas. I readily admit that some modest increases in appropriations have been made for agricultural research programs. Even the proposed federal budget for 2014 shows a comparable percentage increase. However, compared with other areas of research, the exceedingly modest base of agricultural research budgets does not translate into substantial budget enhancement. The increase in appropriations is certainly not commensurate with the challenge.

To begin with, appropriations follow what is perceived to be important. But then this prompts the follow-up question, why is agriculture not perceived to be important? This issue is discussed more fully in chapter 14. First, incremental increases in information, knowledge, and technology have enabled agriculture to maintain steady though modest growth. This has been made possible because of the strong joint relationship with the states and the close association with industry. In fact, the states, particularly those with strong agricultural interest, are major supporters of agricultural research. This joint effort has been sufficiently successful to negate the Malthus concept for 200 years. This has led to a plentiful supply of food—at least for this country and many others around the world, but the threat of insufficient food remains.

Various factors other than information, knowledge, and technology affect agricultural productivity. These other factors include mechanization, transportation, and fiscal support. Also, issues that contribute to agricultural productivity include a better educated and a healthier farm labor workforce. Consequently, success of the agricultural enterprise is not totally dependent upon research, although research clearly establishes the potential and serves as a foundation for success. In this country the issue has gone a step further. There is an expectation of plenty of food and also a high priority on plenty of cheap food. A recent and growing expectation is that food should be sufficiently processed to be oven or pan ready. Agricultural productivity does not appear to be a problem. So why be concerned about research? Consequently, the prevailing attitude is that a problem does not exist, so funding for agricultural research does not need to be strengthened.

Comparison of Federal Appropriations for Research in Selected Federal Agencies

Comparing federal appropriations for research in different areas of investigation is a bit like comparing apples and oranges. Data presented in tables 1 and 2 and figure 14 illustrate that research, in general, has enjoyed substantial federal support. Medical research supported through the National Institutes of Health is particularly robust. The National Science Foundation, which embraces many areas of scientific investigation, has gained substantial support, particularly in recent decades. Appropriations for NIFA, ARS, and ERS have shown generally steady but not spectacular improvement since 1955.

Percentage increases in appropriations during selected periods are even more revealing (table 2). Figure 14 demonstrates that NIH and to a lesser extent NSF have experienced substantial growth in appropriations, while NIFA, ARS, and ERS have experienced only modest increases in federal appropriations during the past half century.

With growth in NIH funding so outstripping appropriation increases in other agencies, it is necessary to use a logarithmic scale to allow growth in other agencies to be seen (figure 15).

In terms of appropriations, agricultural agencies have received only modest increases. In terms of buying power, their funding is nearly flat. In table 3, appropriations from table 1 were adjusted for

Table 2. Percentage Increases in Appropriations for Selected Federal
Agencies for Selected Periods between 1960 and 2010

Period	NIH	NSF	NIFA	ARS	FS	ERS
1960–2000	4362%	2374%	1417%	1116%	1397%	NA[a]
1960–2010	7764%	4233%	2416%	1628%	2008%	NA[a]
1970–2000	1580%	748%	678%	422%	390%	269%
1970–2010	2844%	1386%	1191%	641%	590%	374%
1990–2000	135%	94%	26%	42%	37%	29%
1990–2010	312%	239%	110%	101%	93%	66%

[a] NA = not available.

inflation to constant 1970 dollars using the Bureau of Labor Statistics
Consumer Price Index. These data are plotted in figure 3. Once again
it is necessary to use a logarithmic plot to compare changes in funding among agencies on the same graph, as shown in figure 16. As is
evident in percentage calculations (table 4) and illustrated in figure
4, ARS is nearly level in terms of buying power, and ERS funding has
failed to keep up with inflation.

No increase in federal funding for the ARS and ERS has occurred,
and only very minor increases in NIFA have been made (figure 17).
On the other hand, NIH and NSF have experienced phenomenal
growth in appropriations during this period.

Agricultural research is the most critical research endeavor in terms
of sustaining our country and our civilization. Certainly, a case can
be made that homeland security research seeks to keep us safe. Environmental and conservation research protects and strengthens our
natural resource base and safeguards our environment. Medical research enriches our lives in many ways by prolonging life or providing
a better quality of life. Transportation research helps us to move our
food supplies, other necessities of life, and us as well from one place
to another. Defense research makes the world a safer place in which
to live. However, agricultural investigations that ensure the success of
agriculture constitute the primary research that will help enable our
long-term survival on this planet and preserve our civilization.

Figure 14. Appropriations for selected federal agencies for the period 1955–2010.

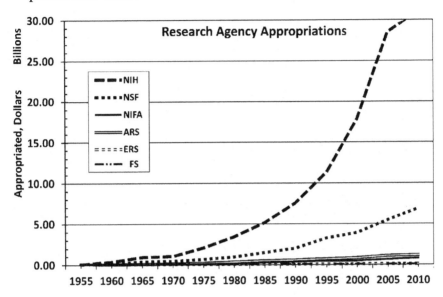

Figure 15. Appropriations for selected federal agencies from the period 1955–2010 (log scale).

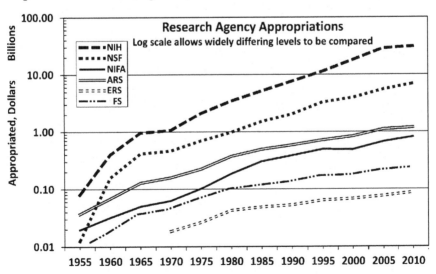

Table 3. Appropriations for Selected Federal Agencies in 5-Year Increments for the Period 1955–2010 in 1970 Dollars (Dollars in Thousands)

CPI Inflation Index	Year	NIH	NSF	NIFA	ARS	FS	ERS
.690	1955	116,161	18,101	28,399	53,575	10,573	–
.762	1960	524,121	208,136	42,167	89,588	19,088	–
.811	1965	1,182,687	512,910	61,651	158,927	43,785	–
1.000	1970	1,061,007	462,490	62,640	159,220	44,424	18,420
1.386	1975	1,510,027	500,094	73,412	161,849	54,403	18,999
2.123	1980	1,615,137	459,317	87,626	175,681	52,535	20,020
2.773	1985	1,856,999	543,480	109,743	177,433	48,497	17,525
3.368	1990	2,249,511	601,562	114,485	173,968	47,261	15,644
3.927	1995	2,876,120	832,765	125,551	181,784	53,776	16,211
4.438	2000	4,015,453	884,038	109,838	187,108	49,085	15,295
5.033	2005	5,661,653	1,088,967	133,038	218,955	55,025	15,060
5.620	2010	5,517,563	1,222,867	143,842	209,900	54,561	15,520

Table 4. Percentage Increases during Selected Periods in 1970 Dollars

Year	NIH	NSF	NIFA	ARS	FS	ERS
1960–2000	666%	325%	160%	109%	157%	NA
1960–2010	953%	488%	241%	134%	186%	NA
1970–2000	278%	91%	75%	18%	105%	−17%
1970–2010	420%	164%	130%	32%	228%	−16%
1990–2000	79%	47%	−4%	8%	4%	−2%
1990–2010	145%	103%	26%	21%	15%	−0.8%

Figure 16. Appropriations for selected federal agencies for the period 1955–2010 in 1970 dollars.

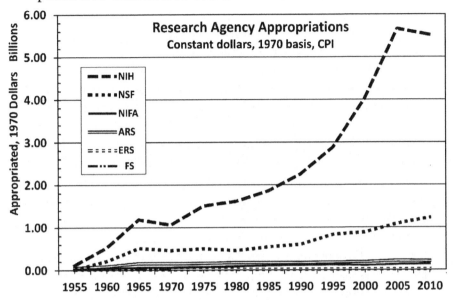

Figure 17. Appropriations for research in 1970 dollars for selected federal agencies for the period 1955–2010 in constant (1970) dollars (CPI, shown as log scale).

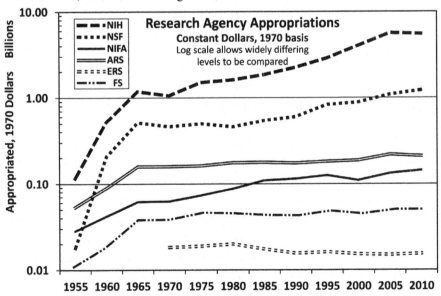

Though, from the historical perspective, a substantial loss in support for agricultural research has occurred, the continued erosion of support in recent years is cause for great alarm. A comparison of federal support in constant 2008 dollars for research and development in recent years (1990–2009) is presented in figure 18.

Agricultural research should be high on everyone's list of research priorities. It is difficult to understand why the federal government has abdicated so much of its responsibility for this critical area, which surely works for the common good of the states. Fortunately, most states appear to have a better vision of the requirements for a secure food future than does the federal government. Unfortunately, agricultural research today is barely an afterthought of the federal government. In the late 1800s agricultural research, particularly after passage of the 1887 Hatch Act, was the predominant research endeavor supported by the federal government. In 1940 agricultural research accounted for about 39% of the federal R&D budget. By 1978 the percentage had fallen to less than 2%. This compares with defense at 45%, the Department of Energy at 16%, and the Department of Health and Human Services at 12% (Office of Technology Assessment, *Assessment*). The total research and development budget for the Department of Defense in 2013 was $72,572,000,000. This was down from $77,500,000,000 in fiscal year 2011. The Department of Defense is by far the largest sponsor of R&D by the federal government. Today agricultural research accounts for a minuscule part of the total federal R&D budget: less than 1.5%.

One of the most disturbing analyses of the funding situation for agricultural research is presented by Heisey, Wang, and Fuglie ("Public Agricultural Research Spending"), who indicate that future growth in US agriculture may depend critically on public investment in agricultural research. Using statistical models they simulated the way in which expenditures for agricultural research might affect the growth of total future productivity (TFP) in agriculture. Their findings are illustrated in figures 6 and 7. These authors base their projections on three funding scenarios: (1) research support stays constant in today's dollars, (2) research support stays constant in inflation-adjusted dollars (with increases just covering the expected rising costs of doing research), and (3) research support increases each year by 1% more than inflation.

Figure 18. Federal appropriations for selected federal agencies R&D 1990–2009 in 2008 dollars. Billions of Constant 2008 Dollars.

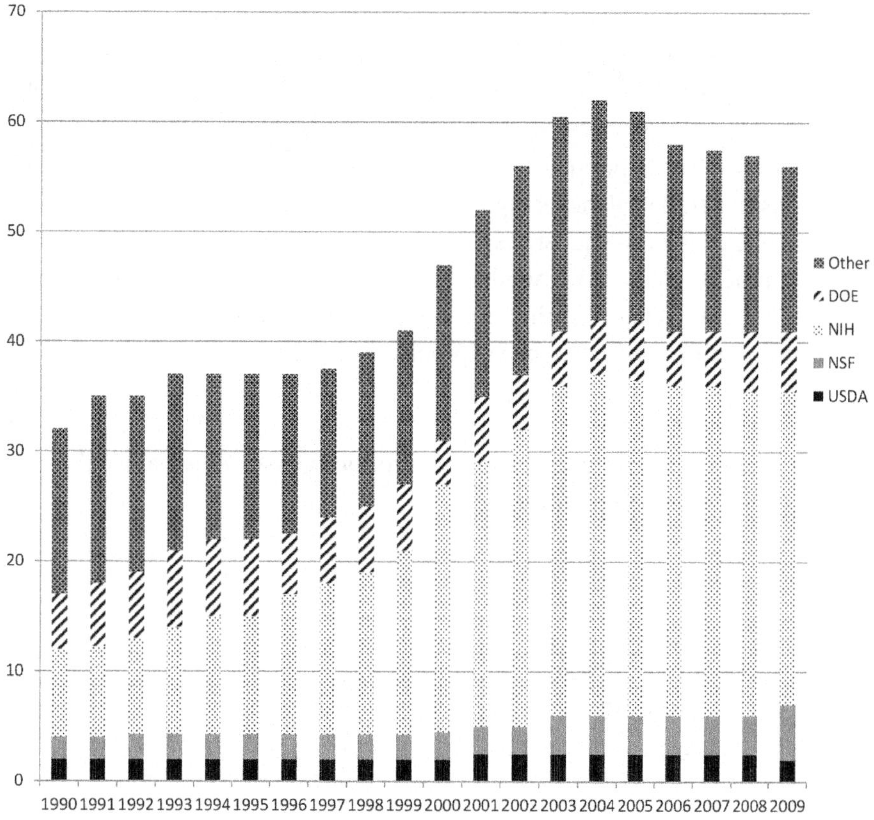

Source: AAAS Reports I–XXXIII.

In scenario 1, real support for research would decrease because of inflation. This takes into account not only the expected rise in general prices but also the fact that research costs tend to rise a bit faster than general inflation. Based on experience, it takes about a 3.7% annual increase to maintain appropriations at a constant rate in real dollars. In scenario 3, it would probably take at least a 4.7% per year increase in minimal dollars to provide a 1% per year increase above the rate of inflation. Under scenario 1, TFP would increase by about 40% between 2010 and 2050 (i.e., US agricultural output would grow by about 40%, using existing land, labor, capital, and material resources).

This is roughly the degree to which the US population, and thus the domestic demand for food, is expected to rise. Under scenario 3, US TFP (or output from today's resources) would grow by about 80% over the next 40 years. This is about the rate that global demand for agricultural commodities is expected to rise. Thus, with declining real spending on public agricultural research, the United States will still be able to feed its population, but its international competitiveness and positive trade balance in agriculture will gradually erode. Agriculture is one of the few industries (along with aerospace activity and entertainment) where the United States has consistently maintained a positive trade balance. Losing our agricultural trade surplus will further undermine the long-standing problems of a negative trade balance and

Figure 19. Possible scenarios for productivity-oriented agricultural research expenditures.

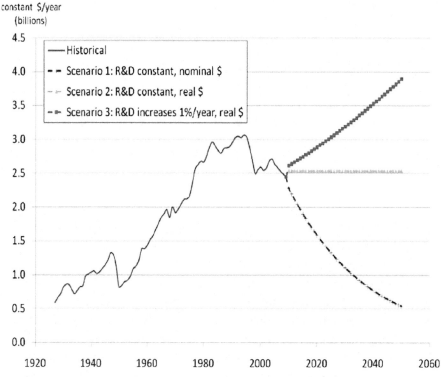

Huffman, "Measuring Public Agricultural Research Capital"; Heisey, Wang, and Fuglie, "Public Agricultural Research Spending."

Figure 20. Projections of agricultural productivity through 2050 (TFP index with 2010 = 100 base year).

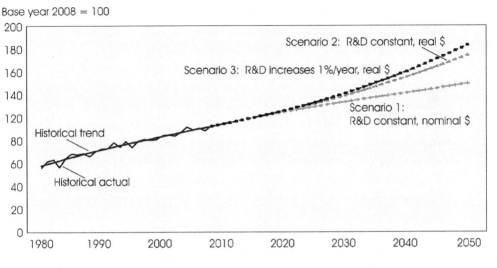

Heisey, Wang, and Fuglie, "Public Agricultural Research Spending."

the country's growing foreign debt. It is abundantly clear that scenario 3 is the preferred approach; scenario 2 is better than scenario 1.

The appropriation trends show that agriculture continues to get a diminishing slice of the federal research pie. While a strong case can be made for many types of public-funded research, an especially strong case can be made for agricultural research. It supports the entire agricultural enterprise, which provides the basic food that nourishes us and supports our health and welfare. These facts provide a sound basis for concluding that, from a national standpoint, successful agriculture ensures food security and contributes to independence.

Funding Mechanisms for Agricultural Research

The history of federal appropriations for agricultural research has been interesting; some might even say colorful. From the very first federal appropriation of $15,000 per station, numerous twists and turns have cropped up. The idea of support for research was growing among the states when the Hatch Act of 1887 was enacted into law. Even with many funding challenges in agricultural research, it is remarkable how successful the system has become. Accounting for much of this success are the highly dedicated agricultural scientists who have been creative in cultivating unique funding sources and developing collaborative relationships. One is left to wonder how success would have been enhanced had a greater level of support been in evidence.

When the Hatch Act was passed, the most reasonable approach to supporting agricultural research was considered what we call today a "block grant," or formula funds, to each state. How the states dealt with the allocation led to a somewhat controversial situation, as discussed in earlier chapters. Congress surmised that the experiment station director in each state knew the problems that needed to be addressed better than anyone in Washington, so the state directors were given control of the appropriation and therefore could determine how the money was to be spent.

The advantage of the block-grant approach is that it gives the responsible administrator complete flexibility to manage the research program. It enables the administrator of the state agricultural experiment station to effectively lead the research effort. It is interesting to note that the block-grant approach has resurfaced in other federal-state programs, reinforcing the argument that local officials know best how to make the most efficient use of such funds. Another advantage of the block grant is that it provides assurance of funding for research projects such as in animal breeding, where maintenance of a cattle herd is necessary for an extended period. The major disadvantage of this approach is that funds may be allocated to less productive institutions. Despite this disadvantage, this mechanism for allocating the federal appropriation in support of agricultural research has served to strengthen the SAES and accomplish much successful research (see chapter 6).

For the United States, World War II was extremely costly in terms of lives and resources and was absolutely devastating to many other countries. However, a few positive outcomes came to light. One was a genuine, enhanced appreciation for research and how it could help end the war and enrich people's lives. Considerable agricultural research was quietly directed to topics that were supportive of the war effort, including food science and technology, insect and disease control, and specialty crops. Many things contributed to the success of the Allies and a favorable conclusion to hostilities. Research was one of those advantages that is often overlooked. However, research not only had a great impact and strengthened the war effort but also affected every aspect of our existence. Many areas in nuclear energy, materials science, chemistry, the medical sciences, and production of selected crops were focused on supporting the war effort and developed rapidly through war-funded research.

Much of the funding for research during the war had been by direct rather than competitive-bid appropriation in support of the war effort. This included major research challenges such as the Manhattan Project, whose objective was to weaponize a nuclear reaction. Once the war had ended and the climate was favorable for enhanced research in many areas, the key question was how to best fund such experiments. It was then that the competitive process was pressed into service. In most areas of research, knowledgeable leaders (such as existed in agriculture with the state agricultural experiment station directors) were not available. The military had its generals and other leaders who could provide research direction just as the experiment station directors in the SAES could. However, without comparable depth of experience the competitive process called for a panel of knowledgeable individuals to identify both the most appropriate research to be done and those who could carry it out. This process would serve not only to identify worthwhile research but also to draw in new, talented researchers.

For a truly open, competitive funding system any scientist or research team can put forward their proposals to explore a promising avenue of science. The funding agency still provides direction by calling for proposals that address an identified need. Since the receiving team or individual may not have the necessary infrastructure and equipment for research, the agency may need to include sufficient funding to build that infrastructure and support equipment requests.

In the competitive process someone or a group of people must determine what research should be carried out. This is an extremely critical step that contributes substantially to the success of any research effort. Done well, this process strengthens the opportunity for success. Done poorly, the results of the research will be disappointing. Once this step has been accomplished, the next step is to determine

who can most effectively accomplish the research in question. Award-
ing money for research competitively is a powerful concept.

Comparison of Funding Models

The advantage of the competitive model for allocating research
dollars is that it provides equal access to all qualified scientists. The
competitive model provides an excellent means of identifying the
most qualified scientists to address a given problem. It also facilitates
tracking and accountability because all funds are appropriated for a
set of narrowly defined goals.

The challenges (weaknesses) in the competitive process are also obvious. First, who decides what is important? Instead of having a director, as with the state agriculture experiment station director, a committee of selected individuals decides what research needs to be done.[2]
Of course, the question is, Does someone who is located hundreds or
perhaps thousands of miles away know best what research is needed
to address particular site-specific or applied problems? (Huffman and
Evenson, "Do Formula or Competitive Grant Funds"). In the basic sciences this is not a concern, but it is a different story for some aspects
of site-specific research needs in agriculture. Evaluation of proposals is
key to determining who can most effectively accomplish the research.
There is always the challenge of omitting some areas of investigation
that could be productive. Another negative is the cost in scientist time
for the preparation of proposals and for the review process by peer
panels. Admittedly, this system works well for research in the fundamental areas of science. It is less effective in many areas of research
where a strictly applied approach is needed (e.g., crop-variety testing,
cultural practices, soil-test calibration, bull and boar testing). Moreover, the competitive process does not take into account the need for a
sustained developmental base for research. Each competitive grant is
viewed as a self-contained, fully supported research project. However,
to maximize the impact of competitive grants, it is important to ensure
they are compatible with the institution's ongoing research program.
This will ensure that such funding streams contribute to the institution's overall research goals.

Today, most federal research appropriations are distributed in
a competitive or a quasi-competitive process. About the only area

where federal support for research still uses an allocation mechanism other than the competitive process is agriculture and in some aspects of defense. In fact, it is fairly easy to pick a fight among agricultural administrators on the issue of formula versus competitive funding. This ongoing controversy has actually led to definitive research to measure the effectiveness of different funding mechanisms (the results have been published in highly respected journals).

Huffman and Evenson ("Do Formula or Competitive Grant Funds") report that in the state agricultural experiment stations, each unit of Hatch formula funding had a greater impact on local agricultural productivity than did a comparable unit of federal competitive grant funding. They speculate that a misconception exists that federal research fund managers can more effectively identify "winning research" than can a local state experiment station director.

Though the debate rages on about formula versus competitive funding for agricultural research at the federal level, most state support for agricultural research is provided by state legislatures in block grants (formula) to the state agricultural experiment stations. This is certainly not unlike formula funding provided by the federal government. Obviously, most state legislatures have confidence in the agricultural research administrators to use public monies wisely to address important researchable agricultural problems in their state. On occasion, a legislator will either formally or informally indicate a specific area where research is desired.

Allocation of federal formula funds is administered in different ways in the experiment stations. In most experiment stations federal funds are incorporated into the base. However, some experiment station directors develop a request for proposals (RFP) and open competition for all federal funds to all personnel in the station for research in a given area. I consider this just creating busy work for scientists. These directors, along with their staff, should know who is most capable of addressing a particular problem, so why not just allocate the money to those scientists?

In-house research in many federal agencies, including ARS and ERS, is accomplished by funding through the block-grant process. Such research is carried out in major research facilities in ARS regional laboratories in Peoria, IL; Albany, CA; New Orleans, LA; and Wyndmoor, PA.

The crown jewel of the USDA's agricultural research facilities is the nation's Henry A. Wallace Beltsville Agricultural Research Center (BARC), in Beltsville, MD, which is the largest agricultural research complex in the world. This massive, outstanding research center is funded by a block grant. Also, ARS has many branch research laboratories widely distributed across the United States and in four foreign countries. Even though ARS is funded by the block-grant method, it employs an elaborate and definitive evaluation and assessment process for research. This occurs for all new projects and for the continuation of existing projects. It is based on an Office of Scientific Quality Review (OSQR) and uses a series of scientific panels specific to major areas of research. The scientific quality review officer is selected generally for a 2-year appointment and is one of the top research leaders in the agency. Although the OSQR does not award money for projects, the review is quite rigorous. The ARS funds can be expended only on projects that have received approval by the OSQR panel. Similarly, ERS in-house research is accomplished in its facilities in Washington, DC.

Differing expectations have contributed to the disagreement about earmarks for agricultural research. It was intended originally that the SAES director would address specific local needs with the formula appropriation. Over the years most SAES directors simply incorporated the Hatch appropriations into their base.

In addition to block grants to institutions and agencies and competitive grants in open requests for proposals, Congress has often used special grants, which are of two types. One may be called "good" special grants. These are directed funds that provide support for new, well-identified, important research problems. They are available to all who have the designated problem and the capacity to carry out studies. The money is allocated either by formula and/or the competitive process. Earmarks that are slipped in without congressional review are now considered to be inappropriate. But those that are reviewed by Congress (e.g., special grants) are not really earmarks in the traditional sense.

Another form of special grants (congressional earmarks) is more controversial. On the positive side, congressional earmarks are additional funds that are included in the budget through the efforts of one or more congressional representatives in support of research carried out by key constituencies. On the down side, resources allocated in this manner do not necessarily fit into any planning procedure. The rise in congressional earmarks in recent years is to a great extent the result of inadequate funding for ongoing agricultural research programs.

The best experiment station directors or research leaders always find some means of preserving certain unallocated resources to meet unanticipated research challenges or to address an emerging problem. Even in tough budgetary times directors have many strategies at their disposal. One common approach is to plan for major equipment purchases very late in the budget year. If no serious emergencies have come up, the purchase can go forward. Otherwise, the money held for the purchase can be used to address a rapidly emerging problem. The major equipment and upgrade needs can be moved to the agenda for funding the next year.

Some argue that directors should maintain a certain amount of un-committed funds so they can move expeditiously to address emerging problems. However, having been an experiment station director in two states, I am well aware that far more challenges are always waiting to be addressed than there is money available to allocate. Consequent-ly, federal formula money was simply incorporated into the station's budget. Of course, the experiment station experienced a loss of flex-ibility as a result. Ideally, each station should have sufficient resources to maintain a reserve.

When a new problem emerged, farmers' commodity groups often became involved in pushing to address the new difficulty. Since the director did not have the flexibility to move quickly, the farmers or commodity interest urged their lawmakers to request an earmark. Specific agricultural problems became one of the prime areas for con-gressional earmarks. I still recall a phone call early one Monday morn-ing from a staff member of a US senator who got right to the point by asking, "What are you doing in research to address the problem of to-mato spotted wilt virus on peanuts and tobacco?" I indicated we were working on the problem although slowly because of a lack of funds. He said the senator wanted to know what it would take to speed up re-search. I gave a figure, and he said, "Fine. I need a brief description of the proposed research." Thus was born a very successful earmark that lasted for several years. Congressional earmarks disappeared from the USDA budget in 2011.

Having been engaged in agricultural research administration for quite a number of years, I can see both positive and negative aspects of various funding mechanisms, including competitive, formula, and congressional earmarks. While serving as USDA undersecretary for research, education, and economics, I had a congressman comment that the executive budget was nothing but the president's earmarks. Obviously, this congressman was a supporter of congressional earmarks. While the executive budget is the president's budget, most items are included after a lengthy vetting process by a research agency and review by the White House Office of Management and Budget.

I do not question the validity of the Huffman-Evenson 2006 study ("Do Formula or Competitive Grant Funds"). I believe they are absolutely correct with regard to competitive and formula-funded research. A place exists for both means of supporting agricultural research. Until agricultural research is adequately funded, some form of specific earmarks will likely remain. The nature of the researchable problem probably determines which funding approach provides the greatest opportunity for success. Though the competitive process is almost certainly the most desirable funding mechanism for basic or more fundamental areas of science, the block grant, properly administered, is the most effective approach when addressing applied or site-specific problems.

In identifying local, practical problems in agriculture the SAES director and staff are probably the most qualified individuals to direct the relevant experiments. On the other hand, basic research that addresses the "grand" challenges in agricultural research described earlier could most effectively be addressed through the competitive grant or perhaps a third process to be discussed in the final chapter of the book.

In 2000 Huffman and Just ("Setting Efficient Incentives") noted several reasons that movement toward more external-grant (competitive) funding rather than formula/program funding may be counterproductive. These reasons are as follows: (1) Scientist time is taken away from directly productive activities and allocated to proposal writing, evaluation, and signaling activities, (2) cost in the form of a proposed budget rather than research output is compensated, (3) quality proposals rather than quality research output are rewarded even though the two are imperfectly correlated, (4) compensation is determined ex ante, necessarily eliminating quality from the incentive scheme, (5) the riskiness of the research is a penalty unduly exacted on scientists, (6) the best scientists tend to focus on grant-proposal writing, leaving the actual research to scientists with less experience, ability, and/or productivity, and (7) peer-review committees sometimes make erroneous judgments about project potential and frequently impose narrow views on research approaches, thus eliminating the benefits of sampling diversity (ibid.). The scientific community is inherently small, so some bias always enters into the review process and cannot be totally eliminated.

I am proud of the open-mindedness of most agricultural research administrators who have embraced the competitive model. Beginning in 1977, Congress first authorized competitive, merit-based, peer-reviewed funding in the USDA. The Food, Conservation, and Energy Act of 2008 (Farm Bill) further stated that "The managers of the conference committee recognize the numerous benefits of competitive research programs and have supported the expansion of funding for these programs. They encourage the department to make every effort to increase support for competitive programs while maintaining the needs of capacity and infrastructure programs when making budgetary decisions." Most of the agricultural research community strongly supported both the idea and the language.

Old ways die hard, however, even those ideas that have merit. Formula funding for agricultural research long served its purpose. That is why as undersecretary I supported a balanced portfolio (i.e., both formula and competitive models, depending on the nature of the research proposed). To illustrate, federal formula funding for both 1862 and 1890 institutions is substantially less than the support provided by state governments. In addition, a major difference exists in state versus federal formula funding for both 1862 and 1890 institutions. The federal formula provides about 90% of funding for research in the 1890 institutions but only 44% for 1862 institutions (Tegene et al., *Investing in People*). In 2000, the states provided less than 15% of the research support for the 1890 institutions but almost 50% of such support for the 1862 institutions. The Agricultural Research, Extension, and Education Reform Act of 1998 requires states to provide a 50% match of federal formula funds for 1890 research, which will likely increase state support for agricultural research in these institutions.

Although methods of allocating federal appropriations are highly debatable, the die is cast. Accepting the peer-reviewed, competitive approach to funding agricultural research programs is clearly our future. Whether or not the agricultural research community wishes to move in this direction is irrelevant. It does not control the situation. However, if it is to be successful, the agricultural research community must embrace the competitive funding model. It is important that the agricultural community work to improve and make the competitive model more amenable to the agricultural research enterprise.

Financial support of intramural agricultural research in the Agricultural Research Service and the Economic Research Service is exclusively by base funding to each agency. However, each of these agencies uses the funding in very different ways. In ERS, the appropriation is a lump sum for the total ERS program. The agency then submits a budget that outlines spending priorities. Congress and OMB monitor its activities. In addition, the authorizing language and provisions in the farm bill include some mandates for the ERS program. Even with this oversight, the ERS administrator has considerable discretion in conducting the research effort in economics. In ARS, appropriations are allocated to a single account for salaries and expenses, but ARS is accountable to Congress for the expenditures according to their location (more than ninety laboratory locations) and the type of research that is conducted with the funds. Consequently, ARS has somewhat restricted discretion.

Funding Agricultural Research: A Future Outlook

We are left with the question, why has public support for agricultural research diminished? First, emergence and industrialization of the concept of research were almost immediately picked up by agricultural interests. Early adopters quickly saw that research could enhance agricultural productivity. Furthermore, there was sufficient agricultural research to enable agriculture to prosper. However, probably more than any other factor, the recognition of the importance and support of research in other endeavors accounts for the allocation to agriculture of a smaller portion of the federal research budget. In this country, the abundance of food has undoubtedly contributed to the lack of concern about future needs.

It is often overlooked that agricultural research has contributed to a better understanding of science and to an appreciation of the research process. Many aspects of statistical analysis grew out of agricultural research. In many land-grant universities, the first mainframe computers were purchased or contributed to by the agricultural experiment stations.

There is one other reason for the lack of support for agricultural research. Though considerable agricultural research is carried out by the private sector, some would dump all agricultural research onto the private sector. This approach would solve one problem (i.e., limiting federal expenditures), but it would create others and in the final analysis diminish the success of our overall research effort. Our agricultural research system has evolved in a somewhat unique way that includes both strong private and public sectors (described more fully in a later chapter). In general, the public sector focuses its efforts on much broader research challenges than the private sector, often concentrating on topics that the private sector does not find profitable. This includes more basic research to advance our fundamental knowledge of agricultural science and technology, as well as applied research on environmental and conservation issues, human nutrition, cultural practices, and other matters. The basic research component itself often stimulates more private research because it opens up new technological areas for corporate R&D to commercialize. Erosion of support for public R&D would reduce the new technological opportunities available to the private sector and eventually erode the corporate sector's willingness to invest in agricultural research as well. In this way the public and private sectors are quite complementary.

Agricultural research has not kept pace with most other research enterprises supported by the federal government. Meeting the necessities for a successful life and a prosperous civilization requires—indeed demands—a successful agriculture. Maintaining a strong agricultural research program is the key to successful agriculture. The highly successful federal-state partnership, which has produced the most successful agricultural research program in the world, must be nurtured and strengthened to meet the future demands of world population growth and increasing expectations for a better life. In times of budget challenges and obviously abundant supplies of food, the first question—Can we afford to commit scarce resources to agricultural research?—is often asked. My response is that it is precisely during these times that we cannot afford *not* to commit the resources for research. A similar argument could be made regarding energy. While considerable amounts of fossil energy are still available, especially with new recovery techniques, now is the time to be doing the research that will enable a smooth transition to a more sustainable energy future.

The most recent farm bill, the Agricultural Act of 2014, includes an innovative and visionary approach to strengthening agricultural research. This act includes provisions for the creation of the Foundation for Food and Agricultural Research (FFAR), which will serve to leverage public and private resources to enhance agricultural research and development. As pointed out in chapter 12, a strengthening of the public-private partnership in agricultural research has great merit. It has the potential to increase the level of funding for agricultural research and to ensure greater relevance of the research effort. To prime the effort, Congress is providing $200 million to the foundation, which must be matched by nonfederal funds. Though this is encouraging news, Goodwin and Smith ("Theme Overview") state that "The bad news is that the new funding streams are insufficient to redress the chronic market failure and underfunding realities that befall US food and agriculture R&D and are unlikely to reverse the dramatic decline in the United States' share of global public food and agricultural R&D spending, with important adverse consequence[s] for the future productivity of US agriculture."

The board of directors of the FFAR includes representatives from a list of candidates provided by industry and another list provided by the National Academy of Science. From these nominees an outstanding and impressive board of directors has been selected to provide leadership for the foundation. The five ex-officio board members include the secretary of agriculture; the USDA's undersecretary for research, education, and economics and chief scientist; the administrator of the USDA's Agricultural Research Service; the director of the USDA's National Institute of Food and Agriculture; and the director of the National Science Foundation. Each member of the board is a distinguished scientific leader who will be able to provide sound leadership. Implementation of FFAR is clearly a positive development in enhancing agricultural research.

I would like to state emphatically that the goal of this book is not necessarily to enhance funding for agricultural research, although that would be highly desirable. Rather, it is to encourage and help develop a better recognition and appreciation of the role of agriculture in ensuring the future well-being of our civilization by providing food and energy. It is also to emphasize the critically important role that agricultural research plays in the success of agriculture and science. Most of the world's people who do not suffer periods of food insecurity take food for granted. On the other hand, those who do suffer periods of food insecurity are often prevented from reaching their full potential.

We as a civilization have been provided multiple wake-up calls. For the future well-being of our civilization both now and in the future, it is imperative that we recognize and appreciate the importance of agriculture and the critical role of agricultural research. Only when we recognize the problem will we be able to design the best agricultural research system and properly fund it to meet the future needs of people and ensure the well-being of our civilization.

In addition to research factors, nonresearch issues can also have adverse effects on future agricultural productivity. Some of these include failure to protect productive agricultural lands, legislation (particularly regulatory) that restricts productive farm practices, and rules and regulations that mandate unproductive practices. Moreover, many other challenges must be dealt with, including the public's perspective of industrial farming, animal rights, genetically modified organisms, fertilizer as a pollutant, antibiotics in animal feeds or for use as a growth stimulant, water restrictions or allocation, and a host of other issues that can affect the future of our food security. While each of these—and perhaps others as well—can have an effect on our future food supply, addressing each of them through dedicated research will undoubtedly contribute to a satisfactory outcome for humankind.

The United States did not invent agricultural research, but, today, early in the twenty-first century, we are certainly among the world's leaders in this vital endeavor, which contributes to a bright future for all people. However, with leadership comes responsibility. Consequently, it is my goal to encourage a better recognition, appreciation, and understanding of agricultural research by all people and not just those few individuals directly engaged in agricultural research.

Wake-up calls have also been sounded by Nobel laureate Norman Borlaug and other distinguished scientists and humanitarians. In addition, the National Academy of Sciences, the President's Council of Advisors on Science and Technology (Office of Technology Assessment, *Assessment*), and other distinguished groups have made public multiple reports. In the final analysis, those of us who have spent our lives in agricultural research are convinced that such recognition and appreciation of agricultural research will enable our leaders to make the best decisions needed to ensure a bright future for everyone in this country and, to that end, every individual on the planet.

11
Responsibilities and Expectations of Selected Groups for Agricultural Research

You cannot escape the responsibility of tomorrow by evading it today.
—ABRAHAM LINCOLN

It follows that all who benefit from agricultural research in some way (chapter 7) bear some responsibility for the research enterprise. Obviously, some are more directly involved than others; some have greater expectations; and some have direct responsibility, while others desire greater benefit. With responsibility come expectations. Clearly, those who share responsibilities have many and varied expectations of the agricultural research enterprise.

Agricultural Scientists

An appropriate place to start is with those individuals directly involved in conducting agricultural research: agricultural scientists. By far, the overwhelming majority recognize their responsibility and are committed to helping solve real-world problems in agriculture. Agricultural scientists, whether university, government (USDA-ARS, ERS, Forestry), or private sector, have much in common. They often have inordinate expectations of themselves and perhaps their ability. Even with all of the positive aspects of their careers, they must still contend with a variety of concerns and apparent challenges.

For example, many of us fall into the trap of dreaming too small. We tend to develop excessive concern about our current well-being and the immediate future. Often this means that agricultural scientists sometimes place scientific advancement, including tenure, scientific papers, presentations, and so on ahead of solving difficult problems. Of course, the "system" in which most agricultural scientists operate contributes to this tendency. Obviously, it is much easier to see and count the benefits of a published, peer-reviewed paper that might be of modest value but fail to see the value of a well-conducted investigation that, however, yields only negative results or no results at all.

Agricultural scientists must be the ones who push back the conventional status quo for the innovations that will ensure our future. This takes courage and does not always lend itself to a peer-reviewed publication. Expectations ought to be greater for those senior academic and government researchers who have earned a modicum of job security and thus have a responsibility to assume a role in high-risk research. Even less secure scientists can carve out a portion of their research commitment to high-risk research, and they should be encouraged and rewarded by research administrators to do so.

In industry, responsibility for high-risk research generally falls under the purview of those who are working at levels above the agricultural scientist. Since one of the ultimate goals of industrial research is to yield a profit, research quickly becomes intertwined with the fiscal aspects of the corporate entity (Jordan et al., *Leadership in Agriculture*). One of the real challenges for agricultural scientists is to see their effort as part of a larger research enterprise. Agricultural science is a young science as compared to chemistry, mathematics, botany, biology, and other basic sciences. These were all well-defined scientific fields of endeavor when formal agricultural research was initiated in the mid-nineteenth century. In the United States, agricultural science was given a great boost with passage of the 1862 Morrill Act, which gave birth to the land-grant university (see chapter 2). As most of these embryonic institutions started developing, it soon became evident that there was inadequate subject matter to teach in agriculture. Thus was born the agricultural research effort.

Agricultural scientists are, by far, the most important part of agricultural research. Having spent much of my early career in agricultural research, I believe the greatest concern of most agricultural scientists is having the tools, equipment, and all other necessary requirements to conduct a research program. All successful agricultural scientists are self-starters who have confidence in their ability and their mission. To fulfill their responsibility, they must become more adept at defining their accomplishments and contributions to society.

One of the challenges agricultural scientists encounter is the need to engage in more high-risk research. Funding, which is a constant problem, will come only when they make a greater effort to achieve success in addressing the more challenging problems in agriculture. Simply researching nonchallenging problems will ensure little or no growth in research funding. Incremental-type research will justify only incremental levels of increased funding. Many nonagricultural areas of research have a major focus on truly visionary research goals such as proving the existence of the Higgs boson. Agricultural scientists are in the best position to identify comparable challenges in agriculture. I realize the entire system must change. In today's climate, a truly visionary research proposal would probably have difficulty getting funding.

Administrators

Essentially all agricultural scientists deal with many levels of administration, especially in academia (e.g., department heads, experiment station directors, administrative heads of agriculture, provosts, university presidents). Comparable levels of administration with different names exist for agricultural scientists in the federal government. Regardless of the level of administration, however, the responsibilities are basically the same. The first responsibility is to facilitate the agricultural research mission. This responsibility is all inclusive, starting with providing adequate resources and related support, a suitable environment for research, and protection of scientists from unnecessary distractions. These are not small orders but are absolutely necessary for success in agricultural investigations. Other, more detailed responsibilities are also relevant, especially to the intermediate-level administrators. These involve providing a satisfactory environment for close collaboration on ensuring that an individual's research is compatible with the overall mission of the unit. This latter task is particularly relevant with a second, third, and perhaps fourth level of administration.

These levels are also critical in establishing the general mission and specific goals of the research effort. While serving as an agricultural experiment station director, I once had a group of entomologists request my support for a research proposal to conduct a major entomological study in the Galapagos Islands. After quizzing the scientists about how participating in this project would contribute to the mission of the experiment station, I became convinced that even though the idea had scientific merit, it would not enhance that mission. Consequently, I rejected their request.

If they had been successful in securing the grant, I would have had great difficulty explaining to agricultural interests in the state why my entomologists were not working to solve more immediate entomological problems in crops and/or animals in our state. Providing support for scientists is an important task for agricultural research administrators in order to ensure success. Such support includes physical means of research support, and sound advice on technical, philosophical, and political matters. One of the most critical tasks is hiring the right person for the job. Indeed, the most influence senior administrators can have on research programs has to do with hiring scientists because that is when the direction of the research is set.

The administrators at each level must be able to articulate and champion research programs within their purview. This is absolutely necessary if they are to be able to attract funding. When asked about a research project, the administrator must always be ready to provide an update or to discuss the project. Of course, this provides the administrator a perfect opportunity to engage another level of administration, such as the department head or the ARS research leader, to provide greater details about the matter. To be effective, administrators must have considerable knowledge about research that is under way in their areas of responsibility.

Department Head/ARS Research Leader

Arguably, the most important position in the administrative structure in academia is department head.[1] For the USDA's ARS and ERS, this person is the research leader, who has the first level of supervision of and responsibility for research scientists. These persons must be extremely knowledgeable about the research effort and the specific capabilities of scientists in their assigned areas. They are responsible for allocating resources and evaluating research accomplishments, two tasks that involve much more than equally allocating available resources and counting publications. The department head or research leader should be able to effectively evaluate each scientist even without seeing that person's publications. They must be sufficiently knowledgeable about each scientist's ability such that they can organize teams of researchers to address specific issues from a broad perspective. After forming such teams, the department head should be able to counsel them and monitor their progress. In addition, the department head should be in a position to advise agricultural scientists who reach a dead end in their research. Department heads will do well to remember that they are responsible for providing the required support for scientists assigned to their unit. This is a tall order, but such is the role of the first level of research supervision.

One of the most critical times for new agricultural researchers is their first research appointment. It is the first-level supervisor that ensures that new researchers get off to a good start by engaging the specific areas of science described in the commitment to hire. Unfortunately, without strong guidance, new scientists will often start digging deeper into their dissertation topic, which often leads to a dead end.

Experiment Station Director

Throughout much of academia in the land-grant university, the next level of administration is the experiment station director. In other institutions this would be a laboratory director, director of research, or a person with some similar title. At this first level all aspects of agricultural research are brought together. Consequently, the director is charged with representing the total agricultural research portfolio and developing the funding request for the entire research program. The director must be knowledgeable about ongoing research and able to articulate results and accomplishments. The department head is a very important asset to the director by enabling him or her to accomplish assigned duties.

Experiment station directors (laboratory directors) must be familiar with the breadth of agriculture and, more importantly, the limiting factors in production and the acceptable level of quality. This means they never have a downtime. They would be well advised to visit as many agricultural enterprises as possible and talk to everyone who is involved in the many aspects of the agricultural and food industry.

Part of keeping abreast of developments in agriculture includes attending commodity meetings, conferences, and research field days and having one-on-one interactions with farmers and farm leaders. To be fully successful, the director must be or become knowledgeable about all aspects of agriculture within the assigned areas of responsibility. This certainly includes but is not limited to what works and does not work and especially what the limitations or constraints are that would hinder success.

Having a good appreciation of all aspects of agriculture is a requirement; however, it is equally important to know the capabilities of the research staff. Maybe the director does not have to know where they live or the names of their children, but they must be familiar with their research program and their research capabilities. Such information is critical if the director is to be successful in providing the leadership that will ensure that challenges are being addressed in the most effective manner; it is especially critical with regard to seeking funding opportunities.

The director must be aware of other alternative assets that can be employed in certain areas of research. For example, scientists in the typical land-grant university represent many disciplines and can often be enticed to work on an agricultural issue. Oftentimes working with the administrator of other components of the university, a director can uncover alternative means of getting particular research accomplished. The same can be said for other agencies in government. The director can even contract with the private sector if that would offer an opportunity for success.

Certainly, the private sector takes advantage of outsourcing research that can be more effectively done by the public sector. For example, in pesticide development, where site-specific data are important, it is advantageous for the private sector to contract with scientists in the public sector to accomplish such research. Likewise, in certain areas (e.g., pesticides required for metabolic studies) the public sector can outsource to industry.

Another critical responsibility is to provide the necessary tools and equipment to ensure the success of the working scientist. Depending upon the particular institution or organization, this responsibility unfolds in a myriad of ways. However, regardless of the system, one key is to ensure wide distribution and understanding of research findings. This is why communications staff are so important to the director. It should be clearly understood that these persons are not responsible for conveying the success of the research programs. The director has that job. Rather, the communications staff facilitates this task for the director.

Again, depending upon the institution or organization, the director is responsible for resources and thus must be an effective champion at the next level of administration if this effort is to succeed. In systems where the director interfaces with the state legislature, greater opportunities for success exist. Even when this is not allowed, the director should establish contact with all of the appropriate people who can influence agricultural research programs. But in the final analysis, the director must accept the responsibility for providing the necessities to conduct research.

The successful research administrator is highly skilled in assisting scientists in securing grants and contracts that augment in-house resources. As discussed elsewhere in this book, extramural grants can greatly boost research programs. The research administrator is responsible for ensuring that grant-supported research is compatible with the experiment station mission and base funding before approving a grant proposal. Careful adherence to this process can ensure a successful competitive and formula-funded research program. Unfortunately, if grant funding becomes the dominant funding source, the research administrator loses much of the opportunity for directing the research.

A change that occurred during the past half century and is still unfolding is the increasing prevalence of the competitive funding process. While this mechanism has great scientific merit, it leaves a great deal to be desired for funding certain aspects of agricultural research. In the competitive funding mechanism, the scientists become entrepreneurs.

They satisfactorily secure funding for their research programs and, consequently, decide how the money is spent—of course, in accordance with the stipulations of the grant. The director is not in the loop. In essence, as competitive funding becomes more prevalent, the director's effort becomes less relevant. While this is a challenge, it does not negate the director's responsibility but just makes his or her work more challenging.

Without question, the single most important task of the director is the identification, recruitment, and employment of scientists. This is not an easy task, but the difficulty does not negate the importance. Employing the best scientists, equipping them with all necessary resources, and providing them with the most appropriate leadership yields a recipe for success.

University Administrators

Often overlooked as having any responsibility for agricultural research in academia is the university provost. As an emissary of the university president and playing a key role in promotion and tenure, these individuals have a major responsibility in the academic agricultural research process. However, the agricultural research faculty of some universities do not participate in the tenure process.

The entire agricultural research spectrum would benefit if evaluations were focused more on the utility of research results or the impact of the research findings on agriculture. Of course, such evaluation is much more difficult to carry out than counting publications or determining the number of research proposals submitted to a granting agency, but it can be accomplished.

In recent years the promotion and tenure process has unfortunately tended to become more bureaucratic by giving more weight to numerical factors such as grant requests submitted, grants funded, journal publications, presentations at scientific society meetings, and so on while diminishing the impact of lower-level administrator evaluations. The result of this trend is that researchers are opting for lower-risk research, therefore bypassing higher-risk studies with potentially greater payoff.

Others on the president's staff can have a positive impact on research. For example, a university's fiscal offices play a critical role because of the highly diverse funding streams for much of agricultural research. The Office of University Legal Affairs is critical to agricultural research, and the University Governmental Affairs Office is particularly important to agriculture because some aspects of agricultural research are politically charged.

University President

It is almost unheard of to think that the university president has a responsibility and a role to play in agricultural research. But this is a critical office, especially for land-grant institutions. As a leader of such an institution they have a responsibility to foster the land-grant mission. Agricultural programs, including agricultural research, are part of that responsibility, so the university president has a role to play. In doing so, a president should have some knowledge of and appreciation for the land-grant mission and, if not a land-grant institution, at least an appreciation for research. A university president can be a great asset to the agricultural research effort—if this individual develops an appreciation for such programs. In order to do that they must have an appreciation for agriculture in general and agricultural research in particular.

I once had a new university president who obviously wanted to impress an important legislator who had evidenced an interest in agriculture. I was glad to help coordinate such a visit to a peanut farm. The president made quite a favorable impression by engaging the farmer in conversation and actually getting on a John Deere tractor and plowing a few rows of peanuts. That was the last time I ever saw this president in a peanut field or any other field.

A university president has multiple responsibilities, but if they attend all of the football games, most basketball games, and a few other athletic events (as most presidents do), one could reasonably expect them to visit some aspect of agricultural research at least occasionally. I know the cry "I have so many research programs," but agriculture and the mechanical arts (engineering) are fundamental to the land-grant mission.

After my half century in the land-grant system I can count on one hand the number of visits by a university president to learn about agricultural research programs. The most memorable one was by an interim president who had retired from a distinguished major liberal arts university and was just learning about land-grant institutions. In visiting the experiment station and seeing how the agricultural programs really worked, he was as enthusiastic as a kid turned loose in a candy store. In my many years as an agricultural administrator I would have greatly welcomed the university president on some of my rounds. It is still not too late for land-grant university presidents to become engaged. My apologies to those who already are!

Time spent with the university president can be quite helpful in generating support for agricultural research programs. University presidents always carry more clout than a dean or an experiment station director. They can open doors with legislators, industry, and the political community. This is not a one-way street. A university president who knows what is going on in agricultural research can use such information in many ways. For example, in most states economic development is often a major concern. Knowing how agriculture can and does contribute can be a great advantage since research is one of the keys to economic development. The linkage is quite apparent.

Similar arguments could be made for active involvement of the secretary of agriculture in support of agricultural research. Observations over many years indicate to me that secretaries of agriculture are usually more involved in and more knowledgeable of activities in the programs in mission areas of the USDA other than the research, education, and economics. They often get more fired up when a problem emerges that can possibly be solved through research. How unfortunate. About the only time they become real research advocates is when a political issue can be addressed by research.

Stakeholders

Stakeholders are those individuals who have a vested interest in the outcome of research. First, no one is interested in the total portfolio of research under way in a typical experiment station or laboratory. On the other hand, many stakeholders have an interest in some aspects of the typical experiment station research portfolio.

Stakeholders are often very passionate about research programs. They usually have the most to gain or lose from such investigations. Consequently, they must assume the lion's share of the responsibility for supporting these programs. They can exert influence on the research process by becoming knowledgeable about the programs. This means learning the nature of ongoing programs, especially the strengths and weaknesses. Stakeholders must seek the most expedient and effective means of influence. This effort can take many forms. First, they can directly communicate with the leadership of agricultural research programs either individually or collectively. However, the most effective way is to be an active participant in a commodity or trade group. In my roles as both an experiment station director and as undersecretary of agriculture for research, education, and economics, I spent much of my time visiting and listening to commodity and trade groups. They were always especially helpful in defining needs in applied research to solve practical, researchable problems.

One way in which stakeholders and most others with agricultural interests fail is to simply not speak up for agricultural research. Too often they do not take advantage of communicating with and enlightening politicians, agricultural leaders, and others about the importance of agriculture and the role of agricultural research. I personally recall an incident during the preparation of the 2008 Farm Bill. Prior to initiation of work on the legislation, Secretary Mike Johanns held hearings around the country[2] to enable anyone who had an interest to share their views, thoughts, and ideas. I happened to attend the hearing held in Georgia. Not a single voice was raised in support of agricultural research in the proposed farm bill. In the meantime, I was appointed undersecretary for research, education, and economics. In reviewing the comments from other state hearings, I was disappointed to learn that stakeholders around the country had expressed so little interest in agricultural research.

The same situation exists in the political season, when candidates are traveling the countryside, talking about issues, and speaking at political rallies. Such events are available to all citizens. This is the perfect opportunity to tell a candidate what you think. In our system one voice counts for little, but a thousand voices are enough to get a candidate for public office to at least listen to your concern.

Myriad opportunities (e.g., dedications of new or renovated facilities, open houses, field days) exist in which to develop visibility for agricultural research programs; they are excellent venues for interacting effectively with stakeholders. The old cliché "If you don't tell people what you want, you should be satisfied with what you get" is about the way it is.

Commodity/Trade Interests and Associations

Commodity/trade groups and associations usually have highly defined expectations. They want, expect, and often demand research that will contribute to the success of their particular interest or commodity. The combined voice of an interest group or a commodity can be helpful to research administrators. While such groups generally place high priority on solving practical problems, they usually have sufficient vision to support fundamental or basic research designed to solve long-range problems or take advantage of opportunities.

While serving in various administrative positions in agricultural research, I always listened carefully to such groups. They usually had their act together. They knew what they wanted and spoke for a wide audience, but such a broad impact carries major responsibilities. The foregoing paragraph notes the positive aspects of commodity/trade area interests and associations. However, these groups should strive to carry out their responsibilities. They can have a major impact on the political process.

Commodity groups just might be the most challenging groups involved in the research process. They unquestionably have considerable political clout—if they choose to use it. Unfortunately, they sometimes use it on an inappropriate target. I have observed that some members of such groups often attempt to secure increased support for research in their area by simply putting enough pressure on the research administrator to transfer support from a less well-defended area. Such areas are often commodities that do not have a strong support organization or new enterprises. Any long-serving research director knows how each commodity group operates. The key is to work with all of them toward securing backing for the total research portfolio.

One very positive effort that brings together a wide array of interests is the National Coalition of Food and Agricultural Research (NC-FAR). As a stakeholder-led coalition that recognizes the need to encourage and enhance public funding for food and agricultural research, NC-FAR endeavors to support all agricultural research, education, and economics. Its diverse membership comes from commodity and trade organizations, scientific societies, and interested individuals. Having been an individual member since NC-FAR was initiated, I am acutely disappointed that not all leaders of agricultural research programs are members.

State Legislators

Since state governments provide substantial funding for agricultural research, state legislators and, of course, state governors are part of the agricultural research process. In some states, the experiment station director, research directors, or administrative heads (dean, vice president, or chancellor) are charged with interacting with legislators with regard to agricultural issues, whereas in other states they are not. Even though the individual who is designated to interact with the state legislature is a matter of university policy, this person must be knowledgeable about agriculture and agricultural research and must interface with the state legislature on an ongoing basis.

Even when the university president reserves the right and responsibility to personally interact with the legislature, the experiment station director, research director, or administrative head of agricultural programs can also take advantage of participation in the political process, as is the right of every citizen of this country. Nonetheless, these individuals would do well to keep in mind that the only person who can speak for the university or commit the university to anything is the university president.

While serving as the administrative head of agricultural programs at a major land-grant university, I had total freedom to interface with the state general assembly—as long as I coordinated with the president's director of governmental affairs. This was an excellent arrangement. In addition to budget concerns, many issues in which input from the administrative head of agricultural programs would be both helpful and useful often crop up. For instance, state legislation pertaining to agriculture can frequently affect the research programs. A person with intimate knowledge of agriculture and particularly research programs would be far more effective than would a university governmental affairs person. Examples of relevant subjects are the implementation of various state laws on seeds, fertilizer issues, and many of the functions of the state department of agriculture. On certain occasions state laws have been enacted that put academic agricultural leaders in awkward positions or in a position to take the lead for some action more appropriate for a politician. The bottom line is that the research leader or administrative head of agricultural programs should have access to the state political process. The astute university president should have sufficient confidence in the university's agricultural leaders to want them to handle this chore for the administration.

State legislators often need to be reminded that although state government is limited in the ways that it can support and influence agriculture in their state, the single most important way to do so is to provide backing for agricultural research, extension, and education programs. In addition, tax issues, transportation, and environmental and regulatory issues can also affect agriculture.

For some states, this is a highly contentious issue because state funding for agricultural research is in direct competition with other components of education—particularly higher education. Some land-grant universities have failed to recognize the uniqueness of the mission of the land-grant university. The common lament is, "why should agricultural research receive more consideration than other components of the university?" Fortunately, in some states, the legislature recognizes the uniqueness of the land-grant university and funds agricultural research (and often extension) in separate budget lines. This arrangement allows the administrative head of agriculture to work directly with the legislature in budget development. Of course, this can put the administrative head of agriculture in an adversarial position with regard to the university president. The administrative head of agricultural programs must work in synchronization with the president, effectively helping each other. To be at odds with the university president is a recipe for disaster.

State legislatures can and should be strong supporters of agricultural research for the simple reason that a strong agriculture is the key to economic growth and development. State legislative bodies led by the appropriate agricultural committees must be knowledgeable about agricultural research in order to be helpful. Along with the myriad other challenges they face, this is no small task.

US Congress

It has long been recognized that the US Congress has played a leading and dominant role in agriculture and agricultural research. When President James Buchanan vetoed the Morrill Act put forward in 1857, Congress persevered. As a result of its effort, Abraham Lincoln signed the land-grant legislation in 1862. Thus was born the land-grant university. Equally important was the role the Congress played in fostering the initiation of agricultural research. Evidence of continuing interest in agricultural research is the fact that one popular form of congressional earmarks is its support of specific agricultural research projects.

Having talked with many members of Congress, I am convinced there remains great interest in and appreciation of agricultural research. Legislators can see the importance of agriculture and its role in our overall economy. Consequently, the idea of forming a food and agriculture caucus in the US Congress has merit. Such a caucus would provide an opportunity for dialogue on the link between agriculture and a myriad of issues affecting our economy, our country, and even our civilization (e.g., climate change, feeding the hungry, and environmental and health issues). Obviously, support of agricultural research would be a nonpartisan matter.

Members of Congress, including both the House and the Senate, have an interest in the nation's agricultural research effort but usually only as it directly affects their constituents. Although many members are supportive of agriculture, they all, for the most part, are in favor of various farm programs, including direct payments; countercyclical payments (targeting mostly price); the Average Crop Revenue Election (ACRE) program, which targets price and yield; the Supplemental Revenue Assistance (SURE) payments program and ad hoc disaster assistance programs; marketing loans; loan deficiency payments; and many government-subsidized crop insurance programs. Other members have special interests in the multitude of nutritional programs or in rural development, conservation, or some aspect of regulatory affairs. Some members are knowledgeable of each of these programs.

While some members of Congress "support" the agricultural research effort, there are only a few real champions of agricultural research in Congress. A true champion would not only be knowledgeable but would also be an effective critic, counselor, and advisor. It is difficult to be overly critical of members of Congress when one realizes the countless areas they must constantly address. They are expected to be knowledgeable about all areas of science, the arts, and anything else about which a constituent has a concern.

In a fifty-year career in academia I have had many opportunities to interface with members of Congress, and I have met a few who appeared to be truly interested in the agricultural research process. They demonstrated their interest by their penetrating questions and insightful comments. Others showed little interest in or knowledge about agricultural research.

In view of the importance of members of Congress to agricultural research, the immediate question is how one can best communicate with them. From the academic perspective, the Board on Agricultural Assembly of the Association of Public and Land-Grant Universities (APLU), which speaks on behalf of the "system," presents an excellent means of communicating. Since ARS, NIFA, ERS, and the Forest Service are a part of the US government, the secretary of agriculture speaks on behalf of these groups. In addition to these formal mechanisms, individuals from both academia and federal agencies can speak on behalf of particular agricultural research programs. This is greatly facilitated when government officials are requested to appear before a committee or subcommittee. Most scientists are quite adept at wrangling such an invitation.

Even in view of the challenges and concerns outlined in the previous paragraphs, members of Congress have a unique and highly important role to play in the agricultural research process. Legislation (Hatch Act of 1887) enacted by Congress and signed by President Grover Cleveland on March 2, 1887, formally put the agricultural research system in place. Since they are responsible for creating this system, they must still accept some ownership of and responsibility for the system's well-being.

I have often wondered why those members of Congress who represent districts or states with a sizeable agricultural interest are not more interested in agricultural research. Probably it has to do with the simple fact that appropriations for other commodity, conservation, or nutrition programs have a more immediate impact. For a member of the House of Representatives, it is important to see an impact in 2 years; for senators the time frame is 6 years. The full impact of research is measured over many years. It would be more helpful if Congress would be thinking and planning for the long term. Consequently, since Congress is primarily responsible for giving birth to the agricultural research system and since agricultural research is one of the primary factors in the economic well-being of our country, one would expect to see evidence of overwhelming interest in this aspect of their responsibilities. Sadly, we are still looking for more champions of agricultural research to emerge in Congress.

Unfortunately, there is a real challenge for many of our national leaders in fully appreciating agriculture or the agricultural research process. This is a long way from the early days of our country, when Washington, Jefferson, and many other national leaders were highly knowledgeable about many aspects of agriculture. Members of Congress now have almost limitless pressures from countless individual constituents and groups, and many do not have a science background. Very few have had any research experience.

This latter point is particularly troubling. In a world where our future is determined to a great extent by research that creates new information and technology, one would hope that elected officials would have a greater appreciation of the process. An excellent example is the nation's effort to outline a path to achieving energy security. While several false starts have been made, there has been little success in agreeing on the best approach. Our effort should not be to implement alternative energy today; rather, it should first be to develop satisfactory alternatives through research. We are not about to run out of fossil energy, but we are on a collision course with reality. We have the time to do the research to develop alternative sources. Yet, just drilling more or employing new recovery technology is hardly a long-term solution.

Unfortunately, our Congress engaged the concept of paying incentives for noncompetitive technology, which distorts the free market. In the short run subsidies will encourage the use of a technology, but in the long run they work in reverse since less incentive exists to improve the technology. This is just the opposite of developing a realistic research-based approach. Paying incentives to use a noncompetitive technology is counterproductive and will probably delay a satisfactory solution. Always remember: "Do the research first!" We should be focusing our research on developing long-term solutions to problems we face with regard to both food and energy.

Government Officials

This category includes individuals who do not have a direct responsibility for agricultural research yet have a vested interest in a successful outcome of the process. At the state level the first person to come to mind is the state secretary, director, or commissioner of agriculture. These individuals have either constitutional authority or authority delegated by the governor to ensure the well-being of agriculture in their state. With this responsibility one would expect that these individuals would be great champions of agricultural research. Typically, however, they often have other problems to deal with and little time to expend on supporting another system that is not directly within their purview or area of responsibility; they may also have little interest in agricultural research. In the early days of this country these offices showed far greater interest in agricultural research. States would be better served by these offices if these individuals took agricultural research more seriously.

Many government officials often fail to even recognize the agricultural research effort in their own state. In one state in which I have worked, the governing board for higher education ensured that maintenance and operating (M&O) funds were made available for each square foot of space assigned by the university system. Even though agricultural research (and extension) accounted for a substantial part of the system's total square footage, these areas were not provided any M&O resources. Ultimately, this arrangement was reorganized, and slowly this situation has begun to be addressed.

Other state government officials such as those in charge of transportation can sometimes be helpful. In one state where I have worked, transportation officials supported legislation that stipulated that roads at experiment stations be treated as state roads. What great support from an unexpected source! Support of agricultural research by officials at the federal level is critically important. Securing support for agricultural research in the nonresearch areas of agriculture is a goal that can pay great dividends. For example, almost every area of responsibility in the USDA depends upon research. Consequently, it would be highly appropriate for these areas to put in an occasional plug for research. The bottom line is that government officials at all levels should be supporters of agricultural research. Oftentimes agricultural research is somewhat like the library: Everybody expects to use it, but no one expects to have to pay for it.

Nongovernmental Organizations

Nongovernmental organizations (NGOs) play a unique role in our society. Since they have their own resources or know where to get them, they are not beholden to any government. This includes major foundations such as Gates, Noble, Rockefeller, Ford, and many others. An excellent comprehensive treatise on the role of foundations in international agricultural research and development was published in 2010 (Herdt, "People, Institutions, and Technology"). Other such groups include Farm Foundation, Council for Agricultural Science and Technology (CAST), American Council on Science and Health, and Animal Health Institute. Some of these organizations (e.g., CAST) have sanctioned and supported studies and issued publications that strongly support agricultural research (see Buchanan, Herdt, and Tweeten, "Agricultural Productivity Strategies for the Future," and Huffman, Norton, and Tweeten, "Investing in a Better Future").

The Farm Foundation sponsors seminars, webinars, and conferences that support agricultural research. This organization meets twice each year to discuss a myriad of topics pertaining to some aspect of agriculture. Membership comprises representatives from production agriculture, agribusiness, academia, and government. Organizations such as the American Association for the Advancement of Science (AAAS), American Institute of Biological Sciences (AIBS), Sigma Xi, and others offer support for agricultural research. For example, in recent years AAAS has published excellent articles on funding for food, agriculture, and natural resources, as well as other research areas that are supported by the federal government (Mervis, "Research Remains" and "Final 2014 Budget"; Malakoff, "Future Is Flat"). Though each of these organizations has its own agenda, they all benefit in some manner from the agricultural research effort. In light of that, it would be appropriate for them to offer support in some fashion for agricultural research. The real challenge is to identify the sensitive issues that would encourage their backing.

Many ad hoc groups have both benefited and supported agriculture research. In 2009 a group of agricultural, forestry, and conservation leaders began a dialogue that became known as "Solutions from the Land." As the effort further developed, participants who shared varied backgrounds and perspectives worked together to develop a vision for American agriculture, forestry, and conservation that will equip us to meet the multiple expectations from the land in the twenty-first century and beyond. This effort was supported by the United Nations Foundation, Conservation International, the Nature Conservancy, and the Farm Foundation. After lengthy discussions, a report was developed (www.sfldialogue.net) outlining some of the challenges in land management along with a vision for maintaining agriculture and forestry productivity and preserving ecosystem service while minimizing input and environmental impact. Solutions from the Land was recently incorporated as a 501(c)(3) foundation.

Scientific Societies

A myriad of scientific societies relate to agriculture is some way. These organizations have many responsibilities and play a key role in agricultural research. For example, they sponsor meetings to report on research and exchange information on matters of science. Most scientific societies have journals that document in a definitive manner the results of scientific investigations. In recent years, a number of these societies have become more active in supporting research. Clearly, some of the medically related scientific societies were instrumental in advocating substantial increases in funding for the National Institutes of Health. They were able to bring a sense of unity to the biomedical community. While not achieving the success of the biomedical community, agricultural scientific societies have also made concerted efforts. The Animal Science Societies and the Tri-Societies,[3] along with the American Society of Plant Biology, have worked to establish research priorities. Scientific societies have been instrumental in establishing coalitions to support the USDA's research programs, such as the Agriculture and Food Research Initiative (AFRI). Clearly, scientific societies have a special place in the agricultural research effort, and it is anticipated that they will continue to play a crucial role in agricultural research in the future.

Consumers

An argument can be easily made that since consumers are the very ones that depend upon the "fruits" of agriculture, they should be one of the greatest supporters of agriculture and agricultural research. Yet, in much of the developed world the disconnect between the appreciation of agriculture and food is growing. This point was made clear by my granddaughter, Lydia, when she was 4 years old. From our garden at the farm, I selected a bunch of young carrots with all the beautiful leaves still attached. Knowing she loved carrots, I showed her how to wash the soil off and told her I had pulled them from the ground just for her. She exclaimed that she did not like "ground" carrots—she preferred "store" carrots! We can overlook the lack of appreciation of the realities of our food source by a 4-year-old child, but all too often many adults fail to understand where their food comes from and what it takes to ensure a secure food future.

The consumer has many desires; indeed, many actually fall in the category of demands. The consumer expects an abundance of all food products, consistently high quality, longer shelf life, minimum loss in processing and cooking, and, of course, ease of preparation. Above all, the consumer expects food to be inexpensive. One thing all of these desires have in common is that fulfilling them depends upon effective agricultural research in both the public and the private sectors. Consequently, consumers have a giant stake in agricultural research programs.

However, these facts regarding the importance of agricultural research do not necessarily translate into any positive response. As yet we simply have no suitable and effective mechanism to bring consumers together. There is, however, great potential in mobilizing consumers to support agricultural research, especially since they constitute the single largest block of special interest in the agricultural research process. Their expectations already exert great influence on many aspects of research.

The critical question for research managers is, how do you engage consumers? Over the years I have appointed consumer representatives to various advisory boards and committees. You are always on your own since there is no association or consumer organization from which to select qualified representatives. This is why it is important to make every effort to provide educational opportunities to enlighten consumers about science in general and, more critically, agricultural science in particular. This has long been recognized as an important challenge. In fact, more than 40 years ago the Board on Agriculture and Natural Resources of the National Academy of Science recognized the importance of providing information about agricultural science and technology to the news media, government officials, and the consuming public. A result of this recognition was the creation of the organization we now know as Council for Agricultural Science and Technology. The challenge is often exacerbated by the anti-intellectual bias of the advertising medium. A simplified description of science in many television commercials does not help in the educational process.

The success of agriculture has enabled civilization to prosper. While significant pockets of food insecurity remain around the world, much of the developed world enjoys a remarkable degree of food security. Ironically, this success is one of the primary causes of lack of concern by much of the population today and can be directly traced to a general lack of appreciation for what must happen in the future as demands increase in response to increasing population, growing expectations, and greater demands on agriculture and forestry for meeting the energy and food needs of the planet. But this is only a part of the challenge.

In view of the exceptional demands and expectations of consumers, engaging them in supporting the agricultural research process must remain a high priority. While few consumer societies are able to serve as multipliers, we must make a concerted effort to identify groups that can and will represent consumers. Some of these groups may include 4-H, FFA, gardening clubs, organic groups, slow food and locally grown groups, and others. It is very important to continue enlightening consumers and building agricultural research coalitions whenever possible.

Scientists must become more effective in communicating with consumers. This presents unique challenges since, in a democracy such as ours, decisions are made or greatly influenced by the people. Of course, today the majority of the people are urban and have limited interest in, appreciation for, or knowledge of any aspect of agriculture, including farming, processing of commodities, or distribution of food. Consequently, without an appreciation of the complexities of modern agriculture, consumers are hampered in supporting agriculture or agricultural research.

Success of agriculture is critical to our survival on this planet, and since agricultural research is inextricably linked to agriculture's success, one must conclude that each of us bears some responsibility for supporting this endeavor. Obviously, some have a greater interest than others. For example, those engaged in the agricultural research sectors have both a survival concern and a vested interest in the well-being of our civilization. In both the land-grant university system and the USDA, each level of authority has a specific role to play. Likewise, various groups external to research institutions have a challenging responsibility. The consumer, who is the ultimate beneficiary—and that includes all of us—must be both a supporter of and an advocate for agricultural research.

When everyone has a responsibility, oftentimes no one readily accepts it. Consequently, our challenge today is to try to build an understanding of and an appreciation for agriculture and agricultural research.

12
Publicly and Privately Funded Agricultural Research

Some people regard private enterprise as a predatory tiger to be shot. Others look on it as a cow they can milk. Not enough people see it as a healthy horse, pulling a sturdy wagon.

—WINSTON CHURCHILL

Because of its breadth, agricultural research lends itself to a wide array of support. Even within the two main venues—public sector and private sector—different categories of participants are found in each area. Sometimes public-sector research is thought of as research for the common good. While that is certainly true, private-sector research must also be of value; otherwise, no one would buy it. It is abundantly clear that both publicly and privately funded research are important for agriculture. Each has some unique aspects as well as challenges. However, it is important to keep in mind that both are important and truly complimentary. It is equally true that neither public- nor private-sector research alone can accomplish the total mission of agricultural research.

Public-Sector Research

Public-sector research is conducted by public institutions with public funds (usually generated through taxes). The expectation is that such research findings will be for the public good. Among the primary public institutions engaged in agricultural research are the state agricultural experiment stations, which are associated with the

nation's land-grant institutions. The experiment stations of these institutions (as described in previous chapters) address agricultural research issues in each state and territory of the United States, as well
as fundamental science issues that affect agriculture. These facilities
are well positioned to address local problems that can often be dealt
with by some aspect of applied site-specific research, most of which,
by definition, is short term. It is research designed to tackle an immediate problem. The challenge for agricultural research administrators is to not allow the entire program to be consumed by applied
research but rather to make certain that a portion of the effort focuses
on fundamental, long-term challenges such as those described in earlier chapters.

Agricultural research embedded in a state agricultural experiment
station is often a part of the agricultural college and other closely
related colleges in the typical land-grant university. However, other
components of the land-grant university often contribute to the agricultural research mission as well. Oftentimes scientists in various
colleges who happen to be engaged in a particular area of basic research do not think of their effort as agricultural research. They often
make significant contributions to solving particular problems that
are pertinent to agriculture. Scientists from other disciplines within the university can bring new perspectives and new ideas to the
process. Collaboration between state agricultural experiment station
scientists and USDA-ARS scientists can often result in a synergistic
relationship.

In addition to the land-grant institutions, other public and some
private institutions are involved in agricultural research. Frequently
the research thrusts of these institutions lie in the area of basic science. However, seldom do these institutions attend to local problems that require a practical or an applied solution. Another major
aspect of public-sector agricultural investigations is carried out by
the USDA, which undertakes other research endeavors outside the
agricultural research, education, and economics (REE) mission area.
For instance, the US Forest Service (also part of the USDA), which is
in the Natural Resources and Environment (NRE) mission area, is in
charge of a major forestry research effort.

In addition to the aforementioned components that support agricultural research, other mission areas of the USDA have some research capability, although these are sometimes described by nonresearch titles. For example, predator control is carried out by the Animal and Plant Health Inspection Service (APHIS) in the Marketing and Regulatory Program (MRP) mission area. Plant introduction studies are carried out by the Natural Resource Conservation Service (NRCS) of the Natural Resources and Environment mission area.

Considerable support for agricultural research is provided by other components of the federal government. This is particularly true for fundamental or basic research. For example, many agricultural scientists are supported by funding provided by the National Institutes of Health. Since basic biology is important to both medicine and agriculture, it is relatively easy to see how many agricultural research efforts could be supported by the NIH.

The National Science Foundation, which by law supports all research endeavors with the exception of medicine, is also a prime source of funding for agricultural research. The biology section of NSF is an excellent funding mechanism for many agricultural research problems. For example, financial backing for determining the corn genome was provided through the NSF funding line. The Environmental Protection Agency (EPA) funds some environmentally related research that pertains to agriculture. Since environmental issues are of great concern to agriculture, it is not surprising that some agriculturally related research is funded through the EPA.

The Office of Science in the Department of Energy has a division of biology that funds a number of agricultural experiments. In fact, the DOE's Office of Science and the USDA's National Institute of Food and Agriculture (formerly, CSREES) have worked together on a number of occasions. Examples are the joint solicitation of proposals for specific research projects and several other collaborative ventures, particularly in energy research.

The Department of Defense and other federal agencies and departments also fund some agricultural research. Public-sector agricultural research involves both state and federal support. A common denominator is that support is provided through public funds provided by tax revenues.

Private-Sector Research

Private-sector research, in contrast to public-sector research, is conducted by private organizations or institutions and paid for with private funds. Most private-sector research is carried out by corporate organizations such as those devoted to production, manufacturing, or processing of agricultural goods and commodities. Much such research focuses on the tools of producing agricultural products and commodities, including development of crop seeds and specific genetic lines of livestock, pest-control technologies, specific fertilizer materials, disease-prevention agents, machinery, equipment, and many others.

Considerable private-sector research is directed at product development—particularly processed, easy-to-prepare foods. The modern supermarket has an extensive assortment of quick, easily prepared freezer meals. Such foods readily lend themselves to trademark patent protection. An important part of this effort is highly skilled photographers who can make any product look better than a dish made in the home kitchen. However, there is still great opportunity for enhanced quality of many frozen meals.

A great deal of research is expended in the private sector to support the manufacture of agricultural equipment and supplies. For example, the development of all aspects of farm equipment, particularly a new design or new machines to harvest a nonmechanized crop, requires great thought and research. Recent innovations in farm machinery have included autopiloted equipment that applies seed and chemicals by very precisely employing global positioning system (GPS) technology.

Developing more effective means of processing agricultural commodities requires much ongoing research. The consumer is always interested in new products that are time-savers with minimum loss in quality.

Private-sector research must lend itself to capturing a profit and therefore must provide a value proposition to the grower/consumer. The well-developed and established patent process in this country is a great asset to privately supported research. It is becoming a more useful tool for publicly funded research as public institutions seek to profit from intellectual property. Some areas of research supported by public institutions provide results that offer opportunities for development into processes or products that can be commercialized. Oftentimes such developments require financial resources that are more readily available from the private sector than from the public sector. In this situation, a cooperative relationship involving both public and private partners can be established that brings benefit to all parties. Bringing publicly and privately funded research together can be a challenge, but it is not impossible. In view of the fact that such collaboration can be highly productive, laws have been enacted that facilitate such an effort while protecting all of the participants.

Yet another important category of private-sector research is carried out by foundations. Philanthropic foundations sometimes conduct experiments in their own laboratories or fund studies in either the private or the public sector. In any event, such research is still owned by the philanthropic foundation to use or dispose of as it sees fit. Since the ultimate goal of philanthropic foundations is to help people in some way, intellectual property is generally used to further the mission of the foundation.

Still another area of private-sector research is experimentation that is carried out or supported by commodity or trade associations. Such studies clearly meet the criteria for private research in that they are funded with private money and the results are owned by the private funder. Research supported by commodity or trade groups is often, indeed usually, complementary to ongoing agricultural research efforts supported by public funds. Consequently, results of such research are immediately available to the funding commodity or trade association. Any resulting intellectual property is usually designated to support the interests of the commodity or trade association. However, public institutions that conduct research for such groups are often allowed to benefit from intellectual property. The distinguishing characteristic of private-sector research is that in addition to private support for the research, the results and the resulting intellectual property are typically owned by the private entity that supported the research to dispose of as it wishes.

Type of Research Carried Out in Public and Private Sectors

The public and private sectors are not constrained with regard to the type of research they can carry out. However, from a practical perspective some generalizations can be made. Since private-sector research requires both a means of capturing a profit and a demonstrated value such that a grower/customer would pay for it, some types of research are more appropriate for the private sector than for the public sector. For example, any technology (e.g., a pesticide or food product) that can easily be patented and can capture a profit if the product delivers desired results might be eligible.

US patent and trademark protection is a powerful means of protecting intellectual property. This is an area where industry excels. Protection of research results (intellectual property) such that they can be used exclusively, licensed, or sold for a fee ensures a return on research. Patent protection, along with various means of extending the life of a patent, provides great *economic incentive* for private research. Research has engendered a wide range of developments such as equipment, production products (e.g., pesticides, soil amendments, fertilizers), and food products that can be protected through the patenting process.

The challenge is to capitalize on the strength of each sector such that the agricultural enterprise is strengthened for the benefit of all. One piece of inspired legislation that helps accomplish this goal is the Bayh-Dole Act (PL 96–517), which, together with later patent and trademark amendments of 1980, created a uniform patent policy among federal agencies that fund research. The true value of this legislation is difficult to comprehend. It enables universities, small businesses, and others to retain title to inventions made even when the research received federal funding.

The patent process offers great protection for innovation resulting from investment in research. In recent years, a patenting process has been developed for plants. It is only through the protection of investigative findings that the supporters of research can realize a profit that enables research to be done in the first place. However, patent protection does not ensure product success; numerous agricultural companies with extensive lists of patents have been driven out of business because their inventions did not improve the grower's profits.

Patent strength combined with grower value can be easily illustrated (for details see Charles, *Lords of the Harvest*). In the fast-paced, developing field of biotechnology, concern emerged as to how to capture a profit from fundamental, basic research. This became a critical issue for the Monsanto Company as it had invested millions of dollars in a particular gene that inactivated one of their widely used herbicides, Roundup (glyphosate). However, Monsanto scientists devised a means of capturing a profit, and the rest is history.[1] The concept of a technology fee was a revolutionary development. Clearly, for a company to invest millions of dollars in research, it must have a means of capturing a profit that justifies the expenditures of funds creating the original development. This concept has led to significant advances in the production of a broad array of food and fiber products.

An often-asked question is, what are the differences in the types of research conducted by the public and the private sectors? The answer is, "in general there are few." The most distinguishing fact is that private-sector research supported by private funding must have a means of capturing a profit to have market success. However, in many cases even this distinction becomes blurred.

In the earliest days of state agricultural experiment stations, new crop varieties were truly for the common good and were available to everyone simply for the asking. However, with protection afforded by plant patents, exclusive marketing arrangements, along with the emergence of university research foundations, it became feasible to capture a profit from development of crop cultivars. When the university rather than the research entity (experiment station) is the assigned owner of intellectual property, a greater effort is often made to maximize profit for the university—even at the farmer's expense. Dealing with such sensitive matters is precisely why university leaders at the highest level should be knowledgeable about and involved in agricultural research. (The role of senior university executives in agricultural research is described more fully in an earlier chapter.) To the farmer, the public institution and private corporations are acting in a similar manner. While this is sometimes a bone of contention, most farmers view paying for particularly useful and productive technology as the cost of doing business. Indeed, this is a benefit to all concerned. If the farmer profits from research, then there is great assurance that research results will be utilized.

In the absence of a profit relationship resulting from public-sector research, new developments sometimes languish on the shelf from lack of interest or innovative marketing. The profit motive ensures an effort will be made to encourage the implementation of new technology. Of course, it goes without saying that only those aspects of research that result in a marketable outcome (e.g., new crop varieties, new equipment) lend themselves to this outcome. However, the lack of an easily captured profit does not mean the research does not have value.

Very often public-sector research does not provide results that can easily be marketed for a profit. Research in areas such as soil fertility, animal grazing, most cultural practices, soil-test calibration, variety testing, and crop rotation fits this category. In addition, public-sector researchers do not bear the burden of market forces because tax dollars are used to fund the work, and any technology created is not required to actually be marketed (although there is a desire to do so). In contrast, private-sector research that does not create products of value will be viewed as a failure, which is why the number of R&D-based companies involved in pesticide development has, through consolidation, diminished to a relatively few who have mastered the discovery, development, and sustained performance of their product portfolios.

During the past three-quarters of a century a distinct effort has been made to move to the private sector research that results in findings that can capture a profit. Certainly, development of most agricultural chemicals for pest control fits in this category. Also, developments in farm machinery and equipment have also moved to the private sector. In addition, a trend has arisen for crop varieties, particularly those developments that incorporate proprietary genetic information, to be developed by the private sector. In fact, many agricultural products companies have solved the problem by integrating with seed companies and using the seed as the delivery system for technologies such as seed treatments and traits. I see this trend continuing.

Employment of the tools of biotechnology in developing useful germplasm that can be incorporated into crop cultivars will undoubtedly become more important in the future. Using techniques such as gene sequencing, molecular markers, and so on helps speed up the process while at the same time creating more useful lines of germplasm. Future food security will require the tools of biotechnology along with the expertise found in more traditional disciplines, including agronomy, plant physiology, entomology, and plant pathology.

In a capitalistic economy, it is logical for those things that can be done by the private sector to be accomplished in that manner. Moreover, the private sector is an engine for innovation in agricultural productivity and will be at the center of the effort to fulfill future food demands globally.

Just as there is a spilling over of public-sector research findings, the reverse is also true. While privately sponsored research is directed at developing proprietary information for use in a saleable product, oftentimes basic scientific principles are developed. Such information can be used for the common good just as if it had been developed by public-sector research. This discussion invariably brings up the question, why not simply let the private sector handle all agricultural research? Clearly, industry does an exemplary job of agricultural research in those areas where a profit can be captured. It is not as successful in other areas, however. On the other hand, the public sector, along with the private sector (foundations), is probably our best hope for addressing long-range research challenges such as described in chapter 9.

Research that affects production agriculture is highly important; however, it represents only one aspect of the investigative efforts needed to enhance the agricultural enterprise. Much research is needed in processing and many aspects of nutrition, which are all critical components of the food chain.

Keith Fuglie and his coworkers ("Contribution of Private Industry") have addressed the issue of public versus private funding for agricultural research and development. They point out that by increasing private-sector funding and decreasing public support for research and development, the question arises as to whether private research and development can substitute for public research and development. They note that if this occurs, productivity growth in agriculture can be maintained. They find that public and private research complement and in fact depend on each other. Breakthroughs from public science stimulate private investment in R&D as private firms seek to commercialize these scientific advances into products for farmers. The private sector also relies on public institutions to train the future scientific workforce, whose work is accomplished almost exclusively in the public sector. An implication of their findings is that if resources for public agricultural research diminish, then private R&D will also eventually stagnate as well, as fewer scientific breakthroughs are available for commercialization and scientific workforce needs go unmet. They also raise the question of a potential problem of private agricultural input industries becoming overly concentrated. The fact remains that much of the basic research in agriculture that is done in the public sector finds wide acceptance and use throughout both the public and the private sector. In a consideration of public and private research one point stands out above all others: Both public- and private-sector agricultural studies are important and highly complementary. Neither can do the job alone.

Funding for Public- and Private-Sector Research

Since both public and private agricultural investigations are important, a particularly relevant question is, what level of funds is appropriate for each sector? Equally important is, who decides the amount of the expenditure? While the answer is not always immediate, one thing is known for sure: The scientists who do the research are not the ones who determine the funds available.

The "who" in deciding funding level is a crucial question. In the private sector, funding for research is simply considered an investment and always has a business case—you spend money to make more money! Not really complicated in principle—but in reality it is exceedingly complex and highly risky. The first steps the private sector must take are to set a goal and then decide whether it is achievable and whether the research results can be marketed. Second, it must figure out whether the venture can turn a profit. Finally, it must come to grips with the question of whether a product can be made and, if so, whether it can be sold at a price the grower will pay.

Marketability and profitability are distinct matters. While managers may be knowledgeable about business, they may not have much appreciation for the science, so they must trust their scientists to do their homework. In the final analysis, the business leaders must often respond to some extent with an instinctive gut response. This becomes relevant when the investment in a new technology is still on the drawing board and may take more than a decade to bring to market. They must decide not only how much of the corporate resources to invest but also how much time will be needed to realize a profit. For these reasons, R&D-based corporations in agriculture are structured in such a way that R&D, commercial, regulatory, and manufacturing functions are all full and equal partners in business development. The independence of all areas is central to success, and a commercial-only focus will not succeed. Effective resistance-management strategies, for example, are critical to the long-term viability of products, and these are developed for all products whether or not regulatory requirements are in place. Effective resistance-management strategies, for example, are critical to the long-term viability of pesticides. It is a general practice to give some thought as to how to ensure that a given product will remain effective in the future.

Management strategies to maintain viability of a pesticide may be influenced by the nature of the pesticide's mode of action. One of the classic examples of this point is the identification and development of the Roundup Ready gene by Monsanto. This approach took a very creative path in that the product Roundup was not changed; rather, the target was changed (see Charles, *Lords of the Harvest*, and Jordan et al., *Leadership in Agriculture*).

The "who" in the public sector is much less clear. In the private sector the scoring is well done by keeping up with profits. On the other hand, research for the public good is not so well defined, and benefits are not always apparent. However, as described in chapter 8, many definitive studies have been conducted that clearly show the return on investment (ROI) for publicly funded research. In recent years a concerted effort has been made by the experiment station system to relate the impact of research to those who are responsible for funding public agricultural research and consumers.

While such information is an excellent means of sharing the value of agricultural research, the support for it is determined by those who control the public purse. State legislators and Congress control all expenditures of public money. The response of these bodies is often based on political interests rather than scientific merit. Because they appropriate money for all state or federal functions, appropriations for agricultural research become problematical. As pointed out in chapter 10, federal support for agricultural research has shrunk from about 40% of the federal support of all research in 1940 to far less today. By 1991, USDA expenditures for R&D were less than 2% of all federal R&D spending ($1.2 billion of $61.3 billion) (Fuglie et al., "Agricultural Research and Development"). This trend continues, and today the percentage is even less. Fuglie and coworkers further point out that about 4% of federal support for research at universities and colleges was for agriculture. The shrinking federal support for agricultural research has occurred to a great extent because of the growing importance to the public of research in other areas, particularly defense, health, and the environment, as well as the lack of appreciation of the importance of agricultural research. Beginning in the late 1970s, support for private-sector agricultural research became greater than that for public-funded studies. This is reflected by the plateauing of public-sector research since 1980, while privately funded research continues to increase (ibid.) (figure 21).

A snapshot of public- versus private-sector funding for agricultural research for 2009 is presented in figure 22 (Economic Research Service, *Agricultural Science Policy*). Total public-sector support (federal and state) was $4.4 billion, compared to $9.5 billion for private-sector funding. By far the lion's share of private funding, $8.7 billion, is for research done by companies themselves (about half of this for agriculture and half for food manufacturing). The remainder is provided by various foundations, commodity and trade associations, and for-profit companies to support research at public institutions. These funds amounted to $855 million in 2009. On the public side, both the federal government and the states are supporters of agricultural research. Of the $3 billion of federal support for agricultural research, about three-quarters is administered by the USDA,

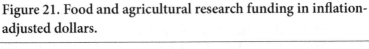

Figure 21. Food and agricultural research funding in inflation-adjusted dollars.

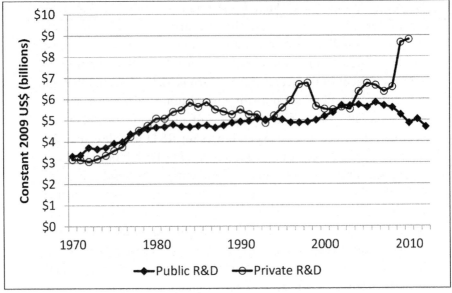

Source: Economic Research Service, USDA, *Agricultural Science Policy*.

with the remainder coming from other federal agencies, including the NIH, NSF, EPA, and others. Just under half of all federal funds for agricultural research goes to the state land-grant colleges and agricultural experiment stations, and half goes to support USDA research agencies. State legislatures appropriated another $1.4 billion for agricultural research in 2009.

Agricultural economists in the USDA Economic Research Service have in recent years made a concerted effort to collect information to develop a better understanding of private-sector and public-sector contribution to agricultural innovation (Fuglie et al., "Agricultural Research and Development"; King et al., "Complementary Roles"). Even though private-sector information is difficult to collect, we have ample evidence that agricultural research and development by the private sector has grown faster and has become larger than food and agricultural research and development spending by the public sector. The growing capacity of private-sector agricultural research and development will exert a great influence on the nature of the research conducted in public institutions.

Figure 22. The US Food and Agricultural Research System.

Funders and performers of U.S. food and agricultural research in 2009

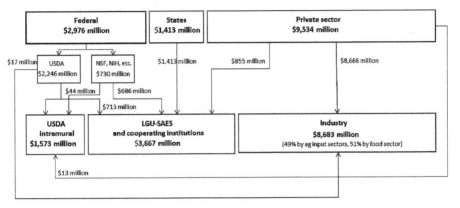

Source: Economic Research Service, USDA, *Agricultural Science Policy.*

The Nature of Public- and Private-Sector Research

Obviously, there are no definitive descriptions of what can be accomplished in either category except in a general way. As pointed out earlier in this chapter, privately funded research must provide a means of capturing a profit, but, in turn, the farmer using the technology must also find it profitable. Research that leads to a product such as a unique piece of equipment, a pesticide, or a tool that enables the user to accomplish a specific task is easily packaged for sale. On the other hand, research that shows that a unique application technique can get comparable results with less of a product will not necessarily lead to a profit for the industry.

However, enhancing efficiency and effectiveness while minimizing use of a crop protectant is a major scientific advancement. Such research with positive outcomes has not only strengthened agriculture but also had a positive effect on human health and the environment. In many areas of agricultural research, successful findings do not lend themselves to capturing a profit. Research that contributes to achieving societal goods such as enhancing quality of life and the environment, improving the economy, and strengthening the community do not, at first glance, appear to result in a means of capturing a profit in dollars and cents. But these research contributions often depend on collaborations among public and private institutions.

Even at the applied-research level, results such as improved crop rotation, better cultural practices, and stronger conservation practices usually do not result in an easily captured profit. In fact, most of the research challenges described in chapter 9 fall into this category.

So far in this chapter, the emphasis has been on the differences between public- and private-sector research. A key question is, what are the values or benefits of collaboration of effort by both the private and the public sectors? First, both sides need to understand their strengths and unique aspects. When this happens, it becomes clear how collaboration can be complementary and strengthen the entire research enterprise. To make such collaboration opportunities into win-win situations, each party should capitalize on its respective strengths.

For example, the development of crop-protection chemicals is a classic example of collaboration strengthening the overall research effort. Industry is well adapted to the synthesis of new chemical compounds and can handle the necessary patenting process. On the other hand, state agricultural experiment stations and the Agricultural Research Service represent an almost unlimited array of opportunities to evaluate a new pesticide under many unique environmental conditions. They are able to conduct field research important to identifying state recommendations, best management practices (BMPs), and integrated pest-management procedures. The strength of industry in synthesis and the strength of the experiment station's diverse, site-specific conditions is a good match that capitalizes on each component's respective strengths. In seeking to develop a collaborative relationship, both the public and the private entity must contribute something that makes the sum of the two components greater than each of them considered separately. If that is not possible, it is wasted effort to force collaboration where no advantage results from cooperation.

13
An International Perspective on Agricultural Research

Yet food is something that is taken for granted by most world leaders despite the fact that much of the population of the world is hungry.

—NORMAN BORLAUG

Even though agricultural research began in Europe, even without air travel or the Internet, the idea quickly crossed the Atlantic and came to North America. Eventually the idea of research devoted to improving agriculture reached into the most remote areas of the planet. It is not surprising that agricultural research has evolved along different lines in various countries. However, reports reflecting a strong relationship between agricultural productivity and research have been consistent and provided the justification for expenditures that have proven to be a worthwhile investment (Alston et al., "Agricultural Research"; Alston et al., "Meta-Analysis of Rates of Return").

Developed societies such as the United States and European countries have long recognized the importance of agricultural research and capitalized on its benefits. Public investment in world agricultural research grew (in 1993 international dollars) from $11.8 billion in 1976 to nearly $21.7 billion in 1995 (Pardey and Beintema, *Slow Magic*, p. 6).

Beintema et al. ("ASTI Global Assessment") point out the difficulties in getting definitive data on expenditures for agricultural research and development. Beintema and Stads ("Measuring Agricultural Research Investments") note that the world is investing less in global agricultural research and development than previously thought (see table). At the time of this study, high-income (developed) countries invested more in public agricultural research and development than did developing countries.

Beintema et al. ("ASTI Global Assessment") also report that global agricultural research and development in both the public and the private sectors increased by 22% between 2000 and 2008. Global public spending on agricultural research and development in 2008 was $31.7 billion (2005 ppp dollars[1]). Spending by China and India accounted for about half of the global increase in R&D in the period 2005–2008. The continent of Africa south of the Sahara accounted for only 5% of growth.

In developed countries public investment in agricultural research grew 0.2% between 1991 and 1996 (cf. 2.2% per year during the 1980s) (Pardey and Beintema, *Slow Magic*). In Africa, they note, no growth at all occurred during this period. On the other hand, China grew 5.5%, while Asia and Pacific countries, excluding China, grew 4.4%. Pardey and Beintema provide substantial evidence of the commitment of various countries to agricultural research.

In some countries, particularly in Africa, where the greatest need for enhanced agricultural outputs exists, support for agricultural research is static or shrinking. Many African agricultural countries spent less than $20 million on agricultural research in 1991 (Pardey et al., "Investments in African Agricultural Research"). Two countries (Kenya and South Africa) spent more than $100 million. In Latin America, four countries spent less than $10 million. Only Brazil and Mexico spent more than $300 million. While growth in research support in some developed countries is not progressing as rapidly as desirable, there is an available base necessary to ensure a reasonable level of ongoing research.

Much of African agriculture is devoted to plants that are not major world crops. These are sometimes referred to as "orphan" crops. To satisfactorily address the needs of many of these countries a committed research effort by the entire international community will be required. Unfortunately, the total research and development expenditures in these countries is only about $1.5 billion annually. Fortunately, a number of philanthropic foundations are leading an effort to foster a greater commitment to agricultural research directed at these countries. A considerable portion of agricultural research is not

Total Public Agricultural Research Expenditures by Income Class and Region, 1981 and 2000

Country Category	Public Agricultural R&D Spending		Regional Share of Global Total	
	1981	2000	1981	2000
	(millions 2005 PPP dollars)		(percent)	
country grouping by income class				
low income (46)	1,410	2,564	9	11
middle income (62)	4,639	7,555	29	32
high income (32)	9,774	13,313	62	57
Total (140)	15,823	23,432	100	100
low- and middle-income countries by region				
Sub-Saharan Africa (45)	1,084	1,239	7	5
China	713	1,891	5	8
India	400	1,301	3	6
Asia-Pacific (26)	1,971	4,758	12	20
Brazil	1,005	1,209	6	5
Latin America and the Caribbean (25)	2,274	2,710	14	12
West Asia and North Africa (12)	720	1,412	5	6
Subtotal (108)	6,049	10,119	38	43

Notes: The number of countries included in the regional totals is shown in parentheses. These estimates exclude Eastern Europe and former Soviet Union countries. Estimation procedures and methodology are described in Pardey et al., Agricultural Research, and various ASTI regional reports at www.asti.cgiar.org. Data are from Beintema and Stads, "Measuring Agricultural Research Investments."

specific to particular sites or regions. This means that research findings developed in one country can often be used in another country.
Clearly, the United States and other developed countries contribute
substantially to the well-being of less developed countries through
their agricultural research effort.

From a global perspective, concern about food security appears to
have diminished over time (Falcon and Naylor, "Rethinking Food Security"). Falcon and Naylor further point out that support was down
one-third from the level only a decade earlier. Fortunately, it appears
that an increasing number of people are developing an appreciation
of the need for greater global food security.

The decision to support public-sector agricultural research and to
determine the level of support needed falls clearly within the purview
of various levels of government in each country. However, in the final
analysis, survival of the human species requires adequate food and fiber supplies made possible through agriculture. Consequently, it is in
the best interest of each nation to nurture and strengthen its agricultural capabilities. One of the prime means of doing so is agricultural
research.

Pardey and Beintema (*Slow Magic*) note some progress in support
for agricultural research in developing countries. In 1990, developing
countries for the first time spent more on agricultural research than
did developed countries. It should be noted that only three developing countries (China, India, and Brazil) accounted for 44% of this
commitment by developing countries. It is indeed unfortunate that
many of the world's countries that have the greatest need to improve
food security are failing to support agricultural research. China, India, and Brazil have made strong commitments to support agricultural R&D (Beintema and Stads, "Measuring Agricultural Research
Investments"). Support for agricultural R&D approximately doubled
in China during the period 2000–2008. Growth in India and Brazil
is not as dramatic as that in China, yet it is still substantial. In recent
years public spending for agricultural R&D in China has surpassed
that in the United States by a substantial amount.

Associated with this greater commitment to support agricultural
R&D is greater agricultural growth as measured by total factor productivity (TFP) (described in chapter 8). The strong commitment to

research enabled Brazil and China to lead the developing countries in the 1980s. Both countries have maintained that rate of growth.

It is unfortunate but understandable that many of the world's countries that have the greatest need to improve food security are failing to support agricultural research. The reasons for this are precisely the same as those that explain a similar trend that is occurring in developed countries, although at a different level. Politicians who make the funding decisions in all areas of government, including agricultural research, must deal with many, many pressures and various needs for financial support. With resources scarce and results from research sometimes in the future (and even then not always guaranteed), financial backing for research is often delayed. Postponing agricultural research usually does not have any immediate negative consequences, which can make that an appealing option for a politician. It takes real

Figure 23. Public agricultural R&D spending by selected countries.

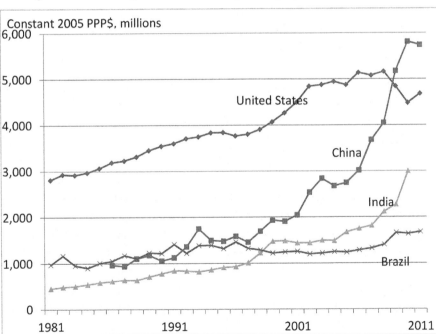

Beintema et al., "ASTI Global Assessment."

Figure 24. Agricultural Total Factor Productivity.

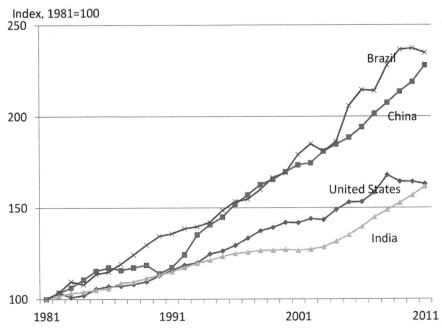

Economic Research Service, *International Agricultural Productivity.*

vision and knowledge to be supportive of investing public monies in research. Unfortunately, some politicians do not have an abundance of vision.

In the absence of adequate resources to carry out all of the studies that would be desirable, it is imperative that we become efficient in using the resources we currently have as we work to develop additional support. Throughout this book I have stressed the need for emphasis on a greater commitment to agricultural research by all countries. Yet, at the turn of the twenty-first century, in many countries support for agricultural research appears to be diminishing. Though some exceptions exist as noted, this phenomenon is occurring in both developed and developing countries. Greater support for agricultural research should be accompanied by more careful prioritization of research efforts and improvement in the efficiency of the research process. It is absolutely imperative that we always spend monies allocated to agricultural research in a wise and productive manner.

Lack of or diminished levels of funding means that those responsible for providing leadership in agricultural investigations must be more creative in leading research efforts. One important way in which this can be accomplished is to develop closer international connections. Most major research universities have a strong commitment to international ties. Leaders of agricultural programs should strive to become intimate players in such efforts. Such relationships are not limited to academia. Various exchange programs supported both by various levels of government and philanthropic organizations provide a means of establishing links. For example, the Consultative Group on International Agricultural Research (CGIAR) is a formal effort to strengthen agricultural research from a global perspective. It involves more than sixty governmental and nongovernmental members and currently supports fifteen research centers and one intergovernmental center.

It is remarkable that this effort laid the foundation of what was to become the Green revolution. The role of the Rockefeller Foundation, working with the Mexican government and involving a world-class scientist, Norman Borlaug, provided the ingredients for an aspect of agricultural research that is nothing short of phenomenal. With the success of the International Maize and Wheat Improvement Center (CIMMYT), the Rockefeller Foundation proposed a worldwide network of such centers. This concept was supported by the Food and Agriculture Organization of the United Nations (FAO), the United Nations Development Programme (UNDP), and the World Bank. These organizations, along with the International Fund for Agricultural Development (IFAD), are current sponsors of CGIAR. This research portfolio, which is directed at some of the most challenging opportunities in agricultural research, focuses on some of the world's most important crops (e.g., rice, wheat, corn, cassava, sorghum, potatoes, chickpeas, and millet). It is also concentrating on other important areas, including livestock farming, conservation of genetic resources, plant nutrition, and water and agricultural policy. Even with its documented success, CGIAR undertook a program of reform in 2001. This is a positive sign for any organization that gives credence to a favorite quote: "If it's not broke, use the time to make it better." This is much more appropriate than "If it's not broke, don't fix it."

The reform effort embraced some new challenges directed at efficiency of water use, improved micronutrient content of food, and employing the power of genetics to create new and better plants. Even with the clearly obvious success of CGIAR, funding support plateaued at about $350–400 million annually between 1992 and 2001. Given the exceptionally high rate of return, it seems apparent there are missed opportunities due to inadequate commitment of resources to support CGIAR. However, this trend was reversed in recent years: In 2011 CGIAR spent $707 million on agricultural research and development (Beintema et al., "ASTI Global Assessment"; Pardey and Beintema, *Slow Magic*).

Importance of Foundations in Agricultural Research

Foundations as a source of support for agricultural research as described earlier are particularly well adapted to playing a role in fostering international research. For many years, the Rockefeller, Ford, Mellon, Kellogg, and other foundations have been active supporters of agricultural research, particularly where it could lead to economic development for less advantaged people.

Foundations have also supported innovative research in this country. An excellent example is the support provided by the Rockefeller Foundation for the aquaculture program at Auburn University. In the early days before aquaculture received significant support from traditional funding sources, the Rockefeller Foundation visualized and backed such research. It saw the potential for this area to be a benefit to this country, and, perhaps more importantly, it saw how the area could benefit people in less developed countries. Following the Rockefeller Foundation, other foundations also supported the Auburn aquaculture program. These included the Kresge Foundation and the Mellon Foundation, whose support enabled the mapping of the genome of catfish. Without the early support of investigative work by these foundations, it is unlikely that the aquaculture industry would have developed as quickly as it did. The financial backing provided by these foundations contributed to the development of the world-class aquaculture program at Auburn University today.

Another effort that lends itself to international cooperation is the Organization for Economic Co-operation and Development (OECD), whose members include some thirty-four countries. The OECD promotes sustainable, sound economic growth and contributes to the expansion of world trade. Agriculture is, of course, indirectly affected by each of these efforts.

Today, foundations such as the Gates, Nobel, Gatsby, and others are making a mark in agricultural research. For example, the Gates Foundation is tackling the challenging problems of improving the photosynthetic efficiencies of rice, which is the staple cereal for about one half of the people on the planet. One of the efforts of the Gates Foundation is directed at enabling C_3 plants to use the C_4 photosynthesis pathway. Some scientists estimate that, if this effort is successful, rice yields could increase as much as 50% (Sage, "Supercharging the Rice Engine"). Success in this effort would truly be a new paradigm in agricultural productivity. This is only one example of the Gates Foundation's commitment to agricultural research. It has many research efforts that focus on some of the challenging problems in agriculture. In addition, much of its work is directed to those parts of the world where food security is problematic.

Another promising effort to strengthen research involves germplasm protections and sharing. By definition this is not research, but it is indirectly related and holds great promise from an international perspective. The National Center for Genetic Resource Preservation is a USDA-ARS facility in Fort Collins, Colorado. The mission of this laboratory is to acquire, evaluate, preserve, and provide a national collection of genetic resources to secure the biological diversity that underpins a sustainable US agricultural economy through diligent stewardship, research, and communication. In addition to this site, the program has several locations around the United States that specialize in selected species of plants. Seed samples are maintained at a considerable cost to the US government and are made available to qualified scientists of all nations free of charge. A comparable policy by all nations would greatly strengthen the total international research effort with only modest expenditures by each nation.

Yet another innovative international effort involving seed is the Svalbard Global Seed Vault, located on the Norwegian island of Spitsbergen, near the town of Longyearbyen (Charles, "A 'Forever' Seed

Bank'"). This effort was initiated in 2006. Planning for the facility and coordinating the shipment of seed samples have been handled primarily by the Global Crop Diversity Trust (GCDT), the Nordic Genetic Resource Center (NordGen), and the Norwegian government. Additional funding comes from various governments and organizations. The facility's location takes advantage of the very cold climate (only 800 miles from the North Pole).

The mission of the Svalvard Global Seed Vault is straightforward. It is to provide a safety net against accidental loss of diversity in traditional gene banks. Such losses of samples could arise from an accident, mismanagement, equipment failure, or natural disaster. By 2010, the seed vault had about half a million samples, with a total capacity of 4 to 5 million samples. Several countries and institutions have contributed samples. The USDA's Agricultural Research Service has contributed more than 20,000 samples.

One of the important opportunities to strengthen the international research effort is personal interchange. Opportunities for graduate training in agricultural subjects at outstanding agricultural universities are one approach. While the trend is often to train students from less developed countries in more developed countries, the reverse can oftentimes be just as advantageous and pay great dividends. Moreover, the exchange of faculty in both short-term and long-term assignments is another way of strengthening international research. Unfortunately, in times of retrenchment caused by shrinking budgets, travel (especially international travel) is often the first casualty. Yet international travel by scientists can be a productive means of enhancing the research effort and should be given very high priority.

Fostering international cooperation in agricultural research is a win-win situation. First, it may help less developed countries become more engaged in agricultural research. Of course, international cooperation brings the strength of different perspectives to any given problem. Finally, and most important, agriculture is a common denominator for all people and all nations. Cooperation is both critical to our very survival and also likely to lead to better understanding among all people.

14
Agricultural Research and the Future of Our Civilization

The future ain't what it used to be.
—YOGI BERRA

Since agriculture's success made possible the emergence of civilization, it follows that the future well-being of our civilization is contingent on the continued success of agriculture. Agriculture was evolving long before there was any form of organized agricultural research. However, the success of research endeavors that began in the mid-nineteenth century greatly stimulated agricultural productivity and enabled the population of the planet to increase dramatically.

As the global population continues to increase and approaches the maximum carrying capacity[1] of the planet, the margin for error gets smaller. Since land is a vital factor in supporting humankind, a country such as India, with a population density in excess of 800 people per square mile, has less flexibility than does a country such as the United States, with only 80 people per square mile. In times of disaster, such as diminished food production or earthquake, more highly populated areas are likely to sustain greater loss of life. Consequently, as the population of the planet increases, we must take greater care to ensure that T. R. Malthus' concern never materializes.

The planet's challenges are daunting. In the past, war and the possibility of a nuclear exchange with some countries gave us all a common topic to worry about. This was especially frightening when families were building fallout shelters, school children were participating in nuclear bomb drills, and the nightly news was continually alluding to the possibility of nuclear disaster. These same fears continue to this day. World events in early 2014 validate the need for continued concern.

You can't build peace on empty stomachs.
—LORD JOHN BOYD ORR

The title of this book and its contents emphasize that agricultural research is among the most important of all research endeavors. This premise in no way diminishes the relevance and importance of other areas of research. However, the key to the survival of all people on this planet is to secure adequate food and energy and ensure that it is affordable to all.

With a more modest planet population of, say, less than two billion, more flexibility would be possible. However, the more the population increases, the less flexibility is possible. Greater population means less room for any miscalculation in food, energy, and water needs. This is because some of the means of agricultural production are fixed or may even be shrinking (e.g., arable land, which is shrinking as more land area is required for housing, transportation, and other needs of an expanding world population).

The land issue is particularly critical. The United States accounts for about 6.1% of the land area of the planet. However, because of its favorable geography, the United States has over 10% of the planet's arable land. Even with this modest land area, the United States produces almost 19% of the world's grains, 22% of all oil seeds, and about 20% of the beef and poultry.

Another critical requirement for agricultural production is water. Obviously, there are many demands on water, one of which is for human needs, but energy-processing techniques that require water constitute another major need. This is especially true for energy sources such as oil sands, tar sands, or, more technically correct, bituminous sands. While these materials contain significant quantities of energy, they require considerable effort and substantial quantities of water to yield a usable energy source. Water is a somewhat flexible resource. Easily accessed water is certainly limited, but in some areas groundwater reserves are replenished by rainfall. In certain situations, desalinization of seawater is a solution—but it is costly.

Many areas of research have well-defined goals and easily recognized importance. These include defense, homeland security, transportation, space, and so on. Of course, the benefits of medical research to individuals and society are quite apparent. However, numerous ailments do not yet have an identified causal agent. On the other hand, many areas of research simply add to humankind's knowledge base. Early studies on optics and planets were not of immediate practical importance at the time, but their findings have proven to be extraordinarily important. Much of medical research has, without question, contributed to a better quality of life for humans; however, some areas of medical research could actually exacerbate the challenge that people face on this planet. Research directed at delaying the senescence of humans could, if successful, lead to increases in demands for food, energy, and water long after people's productive years.

While not as dramatic as a nuclear exchange, adequate food and energy are legitimate concerns that are just as serious. The population story presented by demographers is all too real. Current data and most projections point to a world becoming increasingly overcrowded. This is bad enough, but most of the world's economies are built on an assumption of constant, uninterrupted economic growth. Unfortunately, most nations (including the United States) believe that continued growth is the way to sustain a healthy economy. The two billion more mouths to satisfy in only 40 years is daunting enough, but added to this challenge is the growing expectation not only for more food but also for greater diversity of food, including more fruit, vegetables, and animal products. Thus, these expectations are a daunting challenge.

Energy is nearly as important as food. Both adequate food and energy are necessary for life as we know it. Thinking people who are concerned about the future of our civilization probably put these as number 1 and number 2 as issues for concern. While both are necessary for life, food is a sustainable resource, whereas (fossil-based) energy is a finite resource. Think of these two necessities for life as being stored in large warehouses. The "food warehouse" has two sets of doors, one set of doors for distribution and another for receiving new supplies of food as they are produced. The amount of food coming in reflects the success of agriculture. The "fossil-energy warehouse" has only one set of doors, and that set is for distributing fossil energy. There is no need for a set of doors for receiving new supplies because fossil forms of energy are finite according to our current levels of knowledge and technology, and no new supplies will be forthcoming. However, as pointed out earlier, new recovery technology can increase supplies in the short run, but this does not change the fact that fossil energy is a finite resource.

Research to improve the harvesting of fossil energy such as deepwater drilling, fracking, and drilling in the arctic serves only to enable a speedier consumption of this finite resource. On the other hand, research to foster the success of agriculture enables a quicker replenishment of food for the food warehouse. Another possibility is that some agricultural output can be a source of renewable energy.

Though food and energy may be thought of as the top two, water certainly rounds out the top three concerns. Adequate water is absolutely necessary for the production and processing of food. Likewise, a considerable part of energy reserves are in the form of tar sands. Processing this energy source requires large quantities of water. Since agriculture is a major user of water, much of the water-related research is embedded in the agricultural research enterprise. These last two paragraphs provide a blueprint for the future focus of agricultural research. Food, energy, and water fall under the purview of all agricultural research programs. Consequently, they should receive careful consideration in developing a research agenda.

Though we often think in terms of food, water, and energy as the factors limiting our civilization in the future, the truth is that just about everything it takes to sustain a civilization has limits. In fact, known reserves of many important mineral resources will be depleted in just a few decades (Martenson, *Crash Course*). For the future well-being of our civilization, what must be done? One of the very first responses to this question is to critically and more thoroughly assess the challenges that must be met to ensure future food and energy security. *The information presented in chapter 9 is a good start, but an international symposium of the world's best and brightest should be convened to more fully define the challenges that we face.* This effort in itself presents a challenge. Exactly who should be thinking about the future of the planet? Is it politicians? Hardly! Unfortunately, most politicians are not visionaries but think on a more practical level. They must be concerned about the next election so they can remain a politician. This is not to imply they do not have a role to play; it is just too much to expect them to provide a leadership role. Should it be average people? A good start, but unfortunately many of the world's people are grappling with their own immediate survival. One cannot expect the vast number of people who are struggling to secure nourishment for the day to spend much time thinking about the future. The wealthiest people of the planet have an abundance of everything needed—at a price—for the "good life." They are certainly not concerned about the future. They probably think, "We've always had plenty. Why expect that to change?"

The list of who should be concerned about the food and energy in our future grows shorter! About the only ones left are often individual visionaries, some philanthropic foundations, and the world's universities. This latter group should be leaders in this effort and still can be. All too often, many in our universities are focused on immediate problems or hopelessly irrelevant concerns such as a playoff for a football championship or March Madness. Those of us who have had the luxury of devoting our professional lives to academia have a responsibility to think about and plan for the future—not just the immediate future but also the distant future.

Sustainability of our planet is not the exclusive prerogative of any one organization or group of people. Rather, it is a responsibility we all share. By becoming informed about what is necessary to sustain the people on this planet, we come to the inescapable conclusion that agriculture is critical for our survival. We must recognize that many of the inputs for agricultural production, such as land, nutrients, and water are limited or increasingly costly. The next small step is to appreciate the role of agricultural research. Ergo, agricultural research is our best, indeed, our only hope for civilization.

Then there are the special-interest groups, most of which do not have the commitment of the antiscience, anti-evolution, or antibiotechnology groups. Unfortunately, it is left to a few visionaries such as the late Nobel Laureate Norman Borlaug to challenge us and cause us to think about the future. Moreover, Dr. Borlaug was a great humanitarian who was a skilled leader-diplomat. One of his greatest strengths was his commitment to sound, scientific principles. His strength in science coupled with these other admirable attributes enabled him to contribute greatly to the Green revolution. The World

Food Prize (WFP) was conceived by Dr. Borlaug to honor outstanding individuals who have made vital contributions to improving the quality, quantity, or availability of food for people around the world. Recipients of the prize have come from many different countries and from a range of professional endeavors; however, they all have one thing in common: They have made life better for the people on this planet. While one of the goals of the World Food Prize is the recognition of the food prize recipient, related activities focus on many aspects of food security for all people. The World Food Prize perhaps more than any other activity focuses on the important challenge noted by Dr. Borlaug, who once said, "Civilization as it is known today could not have evolved nor can it survive without an adequate food supply." To recognize Dr. Norman Borlaug's commitment to food security, the ancillary program held in conjunction with the World Food Prize is called the "Borlaug Dialogue."

In addition to addressing current food-security challenges, the WFP Global Youth Institute involves high school students from around the world, two-thirds of whom are young women. This program, held in conjunction with the World Food Prize, provides delegates an opportunity to present research papers on global food-security issues and to discuss science, agriculture, and development with WFP laureates. The institute has influenced more than 1,500 students from thirty-seven states and twenty-five countries.

Knowing precisely what research must be done to ensure success is not easily discernible; however, it is helpful to plan for all possibilities. The agricultural research portfolio is so diverse in this country and around the world that it is difficult to visualize the breadth and magnitude of research needs. Chapter 9 presents a few of the truly grand challenges in agricultural research. Solving any one of these would usher in a new agricultural paradigm. Unfortunately, no one appears to be in a rush to address these kinds of issues. Also, too much of our research effort is devoted to safe, incremental research, which, though not always visionary, has collectively contributed to the most successful agricultural research system in the history of humankind. However, to meet future demands and expectations our agricultural research enterprise should be reviewed critically, rethought, perhaps restructured, and reinvigorated.

Another question discussed at some length centers around determining who is responsible for agricultural research. As pointed out earlier, all who survive on this planet benefit either directly or indirectly from agricultural research. It follows that those who benefit also bear some degree of responsibility. I think we must work toward finding a way to engage each category of beneficiaries in support of agricultural research. This calls into play the vital importance of leadership in all aspects of agriculture. It is refreshing to see many of our land-grant universities including leadership development in the agricultural curriculum. The book *Leadership in Agriculture* by Jordan et al. is a step in the right direction.

Should We Even Be Concerned about Food and Energy Security?

Food and energy are absolutely vital to life as we know it. Until a situation arises that puts their availability in doubt, most people seldom give them any thought. In the United States and many other countries around the world, food security is not a concern for most people. For those who do have a concern, the problem can often be addressed by an infusion of resources. Most countries do not have a shortage of food, but with increasing population, expectations of a better diet, and little hope for solving the infrastructure or resource problem, many people are concerned about our food future. While it is important to enhance production, it is also important that we recognize and address the problem of food waste and spoilage.

Likewise, our concern about energy fluctuates according to several factors. The oil embargo by OPEC countries in the early 1970s was a wakeup call, and concern resurfaced in 2006–2007. Both energy and food security are challenges. Our planet must achieve sustainable food and energy security if our civilization is to survive and thrive. Furthermore, the realization that food and energy security are inseparable is key to our ability to achieve the sustainability of each. A reasonable level of concern can be determined by considering several simple facts. It is possible to assess the status of future energy security. Several facts regarding our energy future illustrate the seriousness of the challenge we face in achieving energy security. Here are ten facts that are not in question:

Fact # 1: World population is growing and expected to exceed 9 billion before 2050.

Fact # 2: The world's appetite for oil is growing and will grow faster in the years ahead.

Fact # 3: Energy consumption by the United States as compared to many other countries will become an increasing challenge for our economy.

Fact # 4: To extend the age of easily recoverable and efficiently processed oil requires new and innovative technology.

Fact # 5: The world will at some point reach "peak oil." This is not a matter of *if* but *when*.

Fact # 6: Many of the highly productive oil fields are maturing according to current technology, particularly in non-OPEC countries.

Fact # 7: The rate of discoveries of petroleum from a global perspective with current technology peaked in the 1960s, although new recovery technology will push this date further into the future.

Fact # 8: Much of the remaining available oil, particularly new discoveries, is located in hard-to-get-at places, such as extremely cold regions and deeper parts of the ocean.

Fact # 9: Much of the available energy that is tied up in tar sands and shale requires considerable water and energy to process.

Fact # 10: All forms of fossil energy (oil, coal, natural gas) are finite.

Acceptance of these facts leads to the inescapable conclusion that we should be as concerned about our energy security as we are about food security. Another reason these two issues are discussed in the final chapter of this book is the fact that each challenge is in the purview of agricultural research.

Food security and energy security are equally important. However, there is one major difference. Food is a sustainable commodity. In contrast, fossil energy is finite. In fact, food availability and the population of the planet have evolved together somewhat. The dire concern pointed out by Malthus has been proven to be unfounded for 200 years because the success of agricultural research has enabled farmers to produce increasingly more food. Such a relationship does not exist for fossil energy. We know how to ensure a secure food future—if we have the will to do it. A secure energy future is more problematic. Though there appears to be some concern about food security, the challenge of energy security seems to be unappreciated. Indeed, television commercials sponsored by the fossil-energy industry appear to be trying to convince viewers that new technology in the recovery of fossil sources of energy can meet our future needs. No doubt new recovery technology will boost the supply of readily available oil and natural gas in the immediate future. Such technology can obviously help meet short-term needs but does not change the long-range scenario. It simply postpones the day of reckoning when the world must convert to a sustainable energy system. According to some, this could be well into the future: "[R]ecoverable resources of shale gas, which when combined with other oil-and-gas resources could last for two centuries" (Nye, "Shale Gas"). The author does not mention a rate of consumption.

The future well-being of our civilization depends on achieving sustainable food and energy security. Concern about both food and energy security can most effectively be alleviated by information, knowledge, and technology generated through research. The best hope for a positive future lies in research.

Who Is Going to Do the Necessary Research, and Where Should It Be Done?

These are the critical questions. The "who" could be thought of in terms of individuals, institutions, economic sectors, or nations. The most logical place to start is by training future agricultural scientists regardless of the venue in which they work. This task falls almost totally on those universities around the world that include the agricultural sciences as part of their curriculum. The important challenge is for universities to recruit "the best and the brightest" for agricultural research. Many of our agricultural colleges' standards for graduate training are not as rigorous as they should be. I marvel at the effort expended to recruit the best football, basketball, and baseball players for the university teams. Once an outstanding individual is identified, every reasonable effort is expended to entice that person to join the team. I would like to see an equal effort in recruiting the agricultural scientists who will shape our future. Again, the age-old question: Who will pay and can we afford it? My response is that we cannot afford to not do it.

The nation's land-grant universities have a primary responsibility for agricultural research programs. However, even these institutions demonstrate a wide range of commitment. Some land-grant institutions have a solid commitment to agricultural research and education, which is clearly evident in the placement of agricultural programs, including research, in the overall structure of the institution. It is apparent that the culture of some land-grant institutions does not place any special emphasis on agricultural programs.

To reflect greater emphasis on agricultural programs, many land-grant institutions place the senior administrators responsible for agriculture (administrative head) at the highest levels of university leadership. A number of land-grant institutions provide for the administrative head to carry the title of vice president. Others provide the title of chancellor or vice chancellor for agricultural programs on campus. The University of Florida actually provides for the administrative head of agricultural programs at the senior vice presidential level. Its administration includes a senior vice president and provost for academic affairs, a senior vice president for agricultural and natural resources, a senior vice president for health affairs, and a senior vice president who serves as the university's chief operating officer. This land-grant institution places agriculture at the highest level within the university and on a par with the medical sciences and other aspects of the university. Louisiana and Tennessee have a chancellor for agricultural programs. Texas A&M University and the University of Missouri provide for a vice chancellor for the administrative head of agricultural programs. A number of institutions (e.g., Arkansas, California, Mississippi, Nebraska, North Dakota, Ohio) provide for a vice president for agricultural programs. Those land-grant institutions that do not choose to place greater emphasis on agriculture provide only a dean title for the administrative head of agricultural programs.

In addition to the land-grant institutions, other state universities and several private institutions have some degree of commitment to agricultural research. These institutions will play an increasingly important role in agricultural research in the future. The next concern is where agricultural research will be done in the future. Both the public and the private sectors have a clearly defined role (see chapter 13). In a capitalistic democracy such as ours, where a minimum level of intrusion by government is valued, as much agricultural research as possible should be accomplished by the private sector. Of course, the profit motive is the driver of this process. Even with a major private-sector commitment, many challenges remain for the public-sector agricultural research effort.

Finally, the agricultural research challenge is sufficiently important that all nations should play a role. As Robert Herdt so eloquently points out, "Every country needs its own people with the capacity to conduct adaptive agricultural research" ("People, Institutions, and Technology"). More developed countries that have longer-established, definitive research programs can provide considerable spillover that benefits developing countries. The agricultural research system that we have in the United States today evolved during the past century and a half. Most will agree this system has served the people, this nation, and the world well. Even though we cannot take credit for originating the concept, we can argue quite successfully that our agricultural research system has been successful and is probably emulated more than any other such system in the world.

However, it is apparent that many people (including me) believe our system must be improved substantially if we are to meet future expectations. This point is strongly validated by the numerous studies, reorganizations, and restructuring efforts of the system that have occurred during the past century and a half. To meet the future expectations of agricultural research and justify substantial funding increases, a thorough review and overhaul of the US agricultural research system is in order. What changes should be made to make this system better? During the development of the 2008 Farm Bill some modest efforts were put forth by two outside groups, and a proposal was made by the USDA administration. Congress accepted none of the proposals in total; rather, it devised yet another structure that accepted a part of each of the three suggestions. Even this effort fell short of making the kind of changes that would really improve the system. We appear to have become expert at seeming to change while hardly changing at all.

If we are to meet the needs and expectations of agriculture in the future, we must develop an adequately funded, more creative, and more visionary agricultural research system. None of the groups involved in the agricultural research process have all the answers, but each one has something to offer. Since everyone benefits in some way from agricultural research, a broad array of knowledgeable individuals should have some voice in the future of this vital process. I am sure that many people would be pleased to offer a completely new design for the agricultural research system. Unfortunately, such an approach would probably not be successful and likely would be counterproductive. Developing an agricultural research system for the future requires the careful input of a wide sector of society. We need input from agricultural scientists and also agricultural science administrators. In addition, we need counsel from farmers, the political establishment, governmental leaders, and (the very important input from) consumers. Each of these categories has a stake in agricultural research and consequently should play a role in determining its future.

Some suggestions for consideration in strengthening the agricultural research system offer a point of departure. This list is by no means complete:

(1) Conduct an in-depth review of the current system to ascertain what aspects of the current system work best and which ones are less effective.

Such a review should not be just another system study. Rather, it should be comprehensive, involving the breadth of those who are intimately involved as well as those who benefit from agricultural research. The goal is to identify the best aspects of the current system so they can be emulated and strengthened. Those aspects of the system that are not successful should be eliminated or targeted for improvement.

(2) Bring all USDA agricultural research into a more comprehensive unit.

A first step in enabling agricultural research to gain greater visibility within the agricultural community is to bring all research in the USDA into a single agency. Congress was probably not aware that it took an important first step in the 2008 Farm Bill by authorizing a chief scientist for the USDA. Consolidating all research and related scientists under the chief scientist would enhance the visibility of agricultural research in the USDA. Sounds simple, but this will be very difficult because no one wants to give up anything for the common good.

(3) Establish an Agricultural Office of Science and Technology (AOST).

To complement the chief scientist position, the research, education, and economics mission area of the USDA should be renamed the Agricultural Office of Science and Technology. This small change would elevate the visibility of agricultural research as a part of the science community in the nation's capital. (This issue becomes moot if recommendation 5 is implemented.)

(4) Institute a means of validating desire and capacity for conducting agricultural research and allocate basic federal (formula) funding commensurate with the total agricultural portfolio and expected probability for research success.

Several years ago a number of agricultural administrators envisioned a reduction in institutions desiring to maintain agricultural programs. A few even predicted that by 2020 only twenty colleges of agriculture would still be in operation in the United States. Though this prediction obviously will miss the mark, a serious question remains as to whether all currently operational agricultural institutions really want to continue to be involved in the agricultural sciences. As institutions evolve, it is likely some will opt out of the agricultural sciences in order to focus on other programs. The relevant question is, Would the overall agricultural research enterprise be strengthened by focusing on only those institutions that truly want to maintain agricultural research programs?

Even among successful land-grant institutions, certain actions call into question institutional commitments to agriculture and agricultural research. For example, recent actions at the University of Georgia provided for the initiation of a College of Engineering.[2] With the creation of the College of Engineering, the Department of Biological and Agricultural Engineering (BAE) ceased to exist as part of the College of Agricultural and Environmental Sciences (CAES). The majority of the BAE faculty were transferred to the new College of Engineering. Some BAE faculty with extension appointments and/or located on the outlying campuses were retained by CAES. These faculty members were given academic homes in other CAES departments.

A validating process would simply provide a means of ascertaining whether an institution continued to have an interest in, a desire for, and a commitment to agricultural research. The challenge is too great and the resources too precious not to spend them on the most capable institutions that also offer the best probability for success.

Though the agricultural research effort grew as part of the nation's land-grant university system, today there are some non-land-grant institutions (e.g., Texas Tech University) that have a solid commitment to agricultural research.

(5) Create a stand-alone agricultural research agency of the US government.

For the past 126 years, agricultural research has been a facet (although a very small one) of the operations of the USDA. At first glance, one would surmise that agricultural research should be in the purview of the USDA. However, upon closer inspection, agricultural research has a much broader venue than agriculture. Today the USDA is, as evidenced by its budget, responsible for the disbursement of funds for an assortment of reasons—all of which are worthwhile and legitimate needs. First is the multitude of farm programs that fulfill a very important function in agriculture and society; the bottom line is that they provide direct benefits to the individual farmer/rancher. Consequently, as any politician knows, things that benefit specific individuals (voters) have a high priority, including the plethora of food and nutrition programs that benefit millions of individuals. At last count, about 47 million US citizens were receiving money from the USDA, courtesy of the US government, from one or more food assistance programs.

Still another important group of USDA programs benefits individuals; these are the conservation programs that are provided to individual landowners for carrying out various and sundry conservation projects.

While there are requirements to receive such monies, the bottom line is that they benefit individuals. In another mission area, millions of dollars go to communities and institutions for water systems, fire stations, and a host of other projects that benefit rural communities. The USDA, along with some other federal agencies, has become involved in venture capitalism by putting up money for supposedly worthwhile projects such as green energy. All of these programs that benefit individuals and communities are not necessarily gifts. Rather, they must be adequately justified, and in the case of farm and conservation programs, many commitments must be fulfilled before an individual receives any money. However, a key is the fact that individuals receive a monetary benefit.

Another major function of the USDA is regulatory. Although pesticide registration and regulation were removed from the USDA and transferred to the Environmental Protection Agency in the 1960s, regulatory responsibility in the Food Safety and Inspection Service (FSIS), Animal and Plant Health Inspection (APHIS) and grain inspection, and packers and stockyards still remains in place. All of these important functions must be funded. Funding for the aforementioned activities constitutes by far the lion's share of the total USDA budget.

The cold, hard fact is that the agricultural research enterprise is far too important to be left to the USDA. This agency has too many responsibilities and pressures that prevent it from providing effective leadership and commitment to agricultural research. The challenges and responsibilities in agricultural research simply have outgrown the USDA. An accumulation of evidence during the past four or five decades supports this contention. Then comes the truly important question, What is the solution? The solution is obvious. One wonders why it has not already been considered.[3] That is, I propose the following: Create a stand-alone Agricultural Research Agency (ARA) along the lines of the National Institutes of Health and the National Science Foundation.

The creation of the National Institute of Food and Agriculture (NIFA), which, in essence, replaced CSREES in the 2008 Farm Bill, was a tacit recognition of the importance of greater acknowledgment of agricultural research. However, NIFA represents only a portion of the public supporters of agricultural research efforts. This legislation has served to focus on one segment of the agricultural research effort. A stand-alone institute that embraces all aspects and components of agricultural research would have a far stronger voice in the science community.

The challenge of such a bold move would be daunting. However, I am confident our present Congress should be able to make the substantial improvements to the current system needed for the benefit of humankind in the twenty-first century and which are critical for survival in the distant future.

In a unified Agricultural Research Agency, the agricultural research administrator would have one of the most important jobs in the government. While the person in this position must have impeccable scientific credentials, it should be a political appointment, but with a specified term of office. While the length of time could be debated, the term should be at least 8–12 years. All other administrators under the ARA administrator should be career appointees who serve at the pleasure of the administration. With the appropriate authority such a person could bring together leaders from various cabinet departments who have some involvement in agriculture. Many of the cabinet-level departments have considerable indirect interest in agriculture as well as some involvement in agricultural research.

(6) Establish a joint cabinet-level council with representation from all cabinet departments under the leadership of the ARA administrator.

At some point in our future, food and energy will dominate the concern of most people on the planet. Long before that time we should be planning how to meet the challenge. Since each cabinet-level department has some role in and responsibility for the agricultural enterprise, it is logical to start now to develop close and effective working relationships. An ARA administrator is the most logical position in the US government to coordinate and lead such an effort.

(7) Identify the research challenges that, if accomplished, would bring about a new paradigm in agricultural productivity.

Chapter 9 identified several challenges that, if accomplished, would have a significant positive impact on the success of the agricultural enterprise. This is only a beginning. Developing some type of venue to bring together leading scientists from various areas of the agricultural and other sciences would be a first step in defining other challenges that will lead us to a more secure food and energy future. While this idea presents many opportunities, one of the most appealing is the nexus of agriculture, nutrition, and health. The connection between agriculture and medicine offers great potential for the benefit of humankind.

(8) Develop more effective mechanisms for capturing research expertise that exists in nonagricultural institutions and nonagricultural components of the government or university.

Successfully bringing together all aspects of agricultural research would go a long way in capturing and fostering the unique capabilities of individuals outside the traditional system. Encouraging such individuals to address an agricultural issue often has unexpected advantages as they bring different and unique approaches to solving a problem.

(9) Strengthen the funding base for agricultural research.

Funding trends in this and many other countries (discussed in chapter 13) provide evidence that the current model is not working. In fact, many of the recommendations mentioned here must be addressed before any effort is expended to strengthen the funding base for agricultural research. It is important to keep in mind that funding for public-sector agricultural research depends upon multiple successes. Consequently, care should be exercised in exacting the maximum support from each area. Over the years, little thought has been given as to how best to cultivate each funding sector. For example, federal support will be more easily justified in addressing problems that are of national concern. Individual states could probably be more successful in more site-specific research. I recognize that these are not hard and fast relationships but merely guides. I also acknowledge that this approach is somewhat at odds with the 1887 Hatch Act legislation. But in all candor, how does one justify providing less funding from the Hatch appropriations to a state such as California, which produces more than 400 commodities, than that provided to a state with a much smaller and less diverse agricultural portfolio? Of course, one could argue that the number and value of commodities are not good measures of needed research commitment. Unfortunately, many commodity and trade groups stress this relationship, especially if their particular commodity generates a sizable farm gate value. Research commitment should be made that will yield the greatest outcome value divided by research expenditure. Again, this is an inexact relationship but in principle much more relevant than basing research expenditures on the farm gate value of a commodity or political clout.

(10) Seek more innovative means and sources of funding agricultural research.

To improve and increase the funding base for agricultural research, a concerted effort must be made to expand all potential sources of financial backing. While state and federal governments provide a substantial part of the funding for agricultural research, greater effort should be directed at other potential sources. Local governments fund very little agricultural research; however, they constitute a major funding source for cooperative extension. Some commodity and trade groups already provide funding for research that is of particular interest. This source has considerable potential for expansion. Another excellent source, industry, already supports much research, mostly specific, proprietary studies. Effort should be directed at encouraging both commodity and industry interests to subsidize more fundamental research in addition to that which directly benefits the commodity or industry group.

Another often overlooked category is specific consumer interest groups. In addition to research funded through taxation (where all of society participates), certain categories of consumers should be encouraged to support both applied and fundamental research that promotes specific areas of interest (e.g., consumers who have special dietary concerns). This model works rather well in the medical field. Some consumer groups are effective in not only energizing their members but also mobilizing the community at large.

(11) Develop a more enlightened philosophy of what constitutes success in research.

Peer-reviewed research papers have been the valued "currency of the scientific realm" for many years. In today's world, they are still important, but we must find additional avenues to measure success in research, particularly agricultural research. We have even fallen into the rut of counting research grant proposals submitted as a measure of research success. Sometimes we even have elaborate systems for assigning a certain weight for senior author, number 2, 3, 4, 10, 20, and even the 100th author! This absurdity has been pushed not by research scientists but by the fiscal accountants or lazy administrators of the world. This matter has led to the coining of the term "least publishable unit" (LPU). The goal is not necessarily to solve researchable problems but rather to maximize the number of publications. This is a quick way to bulk up a résumé. Often high-risk research does not lend itself to as many publishable research accomplishments as do more predictable studies.

(12) Develop more effective international collaboration.

This is a simple statement, but it is often overlooked as a means of strengthening agricultural research. I recall that as a young agricultural researcher about a half century ago, international visitors often came to our department. Many of these scientists readily shared information about their progress while absorbing as much as they could about our programs. For several reasons, I, as a young scientist, had difficulty securing support for out-of-country travel. First is the bias against international travel by public employees. For half a century, I have observed that, with any kind of budget problems, almost always the first action is to restrict or eliminate travel. What a pity. We are among the most mobile people on the planet, yet we have a general aversion to travel by public employees. In addition, sometimes we get the feeling we have all the answers and do not need to learn from others. Again, a pity. Visiting other countries is a first step to developing more collaborative research relationships. Notwithstanding the bias against public travel, I am very pleased to see the agricultural research program at both the federal and the academic levels greatly strengthened by international collaboration. In this area we are moving in the right direction.

(13) Improve funding mechanisms for agricultural research.

It is a bit ironic that an enterprise so important to the well-being of our civilization has never been adequately funded. From the very beginning and throughout history, agricultural research has not been adequately funded, especially in recent years as the emergence of other research venues has exacerbated this problem. Currently two primary funding mechanisms exist that support agricultural research by the federal government. In the USDA, ARS, Forest Service, and ERS are funded through base funding. The National Institute of Food and Agriculture provides funding to the state agricultural experiment stations through both block (formula) funding and a competitive funding mechanism. Managing each of these mechanisms in the most effective manner contributes to the overall success of the research endeavor. The state experiment station directors have great freedom to manage formula money as they see fit. Most directors simply build these funds into the station budget base (as mentioned in an earlier chapter).

A third funding mechanism should be considered for some areas of agricultural research. Major agricultural research challenges require a major funding stream not bound by a formula or competitive model. Here I am visualizing a laboratory devoted to a single mission with a consistent, long-term commitment of funding all the way through to success. This would be more along the lines of funding that industry uses to support research. It is also similar to a model that some foundations employ. Such a system would fund only those laboratories that offer a convincing justification for possible success. Funding would continue commensurate with progress in the assigned research.

This funding mechanism would support really visionary challenges in agricultural research. We should seek challenges that are on a par with finding the Higgs boson! Confirmation of the existence of this particle advances our understanding of the universe. Perhaps not completely analogous but close would be converting C_3 rice into C_4 rice (von Caemmerer et al., "Development of C_4 Rice"). Von Caemmerer and coworkers estimate that it will take 15 years of research for the optimization of a C_4 phenotype and field testing before it gets to farmers (ibid.). Even though this laboratory probably has adequate resources, what if ten such laboratories were working on this same problem? Perhaps favorable results would come sooner. Solving these types of challenges will help secure our food and energy future. I consider this funding approach to be very important for the future.

(14) Develop more effective means of communicating successes and contributions made in agricultural research.

Agricultural communicators have worked hard to publicize research findings to consumers and others who benefit from such efforts. In the future, we must become more proficient at this. It is not enough simply to report research findings; we must also show how our lives are made better through research (or made poorer because of a lack of research). In sharing information about agricultural research we seem to get better and better at a smaller and smaller target. I am very impressed with the job agricultural communicators do. But we have to achieve a new paradigm in agricultural communications. This is not just a job for the communications personnel; rather, senior administrators must be in the lead.

For example, how do we communicate with senior university leaders or key people in a university system, governor's office, or state government agencies? At one time at the state level, governors and their commissioner, director, or secretaries of agriculture were intimately knowledgeable about agriculture in general and agricultural research in particular. We must reestablish that relationship. We must reestablish and strengthen the awareness that agriculture in the United States is a major component of economic development. I readily admit that some governors were excellent leaders and exercised far-reaching vision.

Much of the impetus for developing corn-based ethanol goes to the then governor of Nebraska, Mike Johanns, who, working with his fellow governors, did something to help corn farmers, and the rest is history. Even though debate over the overall efficiency of corn-based ethanol is ongoing, this effort has clearly illustrated the potential for a future sustainable energy economy. Lost in the food-versus-fuel debate is a simple fact: Corn-based ethanol illustrates that agriculture will be a critical component in achieving sustainable energy security in our nation. Agricultural research needs to be on the agenda of the various governors' conferences.

Similar arguments could be made at the national level. If the future of the world depends on agriculture—and it does—the G7 (formerly G8) and G20 assemblies should pay some attention. A tall order, but we must make the effort. Recently, through the dedicated efforts of Catherine Woteki, USDA chief scientist and undersecretary for research, education, and economics, a concept was initiated, and the first meeting with her counterparts was held in conjunction with the 2012 G20 summit in Guadalajara. The first such meeting provided a beginning for international agricultural research leaders to exchange ideas and to discuss the future of agricultural research. Perhaps more important is the fact that such a session brings attention to the significance of agriculture in general and agricultural research in particular.

USDA Chief Scientist Woteki has identified six strategic platforms from which to achieve sustainable food security:

1. Open access to scholarly publications
2. Open access to germplasm collections
3. Open access to genomic and genetic data
4. Accelerated technology transfer
5. Improved global agricultural statistics
6. Establishment of regular coordination of the world's chief agricultural scientists on agricultural research and development in conjunction with the G20 Summit.

These six items constitute an exceedingly sound basis for strengthening agricultural research collaboration.

(15) Improve the quality of leadership of those responsible for agricultural research programs throughout the system.

Almost 25 years ago, the Experiment Station Committee on Organization and Policy (ESCOP) established a Special Initiatives Committee to determine how agricultural research could be strengthened. One outcome of this process and a specific recommendation was the need for enhanced leadership among those who were responsible for agricultural research in the nation's experiment station and federal agricultural research systems. This led to the establishment of the ESCOP Leadership Development Program (Jordan et al., *Leadership in Agriculture*). As Jordan et al. point out, those who have participated in this leadership development program have become more effective leaders in agricultural research (ibid.). To meet the research challenges of the future, it is quite apparent that we must have exceptionally well-qualified leadership throughout the experiment station and agricultural research system.

(16) Encourage more visionary, high-risk research.

A great need exists for more visionary, high-risk research on some of the more challenging problems in agriculture. It seems we are creating an environment that provides greater and more constant rewards for incremental research. Admittedly, it is far safer to just engage in a more conservative research approach. The funding crisis in agricultural research will not be addressed unless we become more visionary and creative and tackle some of the really big challenges in agriculture. For this to occur, much has to change.

(17) Recruit those individuals who are among the best and the brightest.

If we are to solve some of the important problems that will ensure a bright and successful agricultural future, then we must engage the most creative minds to pursue agricultural science studies. This is no small task since more difficult paths of study offer few rewards. A part of the problem is the less than ideal state of our educational system in the country. But agriculture does not even fare well when compared with some other endeavors (e.g., medicine). Many high school guidance counselors have little appreciation for agriculture and certainly not agricultural research. They are often inclined to urge a really bright student to pursue something other than agriculture. So the challenge starts at the beginning—with recruiting the best and the brightest students for agriculture and agricultural research.

(18) Become more critical evaluators of research projects and also do a more effective job of establishing research priorities.

Regardless of funding status, we must become more critical of where and how our resources are expended. First, greater effort should be expended on identifying and developing research priorities. In doing so, we should consider the potential for success and measurable outcomes. Once priorities are established, a reasonable and realistic plan should be developed along with allocation of resources to address the priorities. Know when to "pull the plug" on non- or less productive research. It is imperative we become more critical in evaluations of research effort.

The Way Forward

This book makes a definitive case for the importance of agricultural research for our future well-being.[4] While the significance of such investigations is apparent to those of us engaged in agricultural research, it is equally obvious that today our concern is not universally accepted, appreciated, or even recognized. This is in sharp contrast to the earlier days of our country, when agriculture was held in higher regard. Even though agricultural science is relatively new in comparison to chemistry, botany, mathematics, and so on, support for agricultural experimentation was the primary focus of research initially supported by the federal government. In view of the institutionalization of agricultural research in England 170 years ago and in the United States only 125 years ago, one is left to wonder why the general enthusiasm for agricultural research has waned. Several reasons may account for the loss of enthusiasm and support for agricultural research, but two stand out as particularly relevant.

Historical Abundance of Plenty and the Complex Reasons for It

The American public has confidence in the nation's ability to provide an abundance of relatively safe and cheap food. Indeed, this may be the answer, but is the degree of confidence deserved? How many assumptions must be made in order to support it? This country has always had an abundance of food, so why would anyone expect that to change? However, even though something has always been available in the past, it is not a certainty that it will be available in the future. This is especially true when so many variables exist that cannot be easily controlled.

Closely allied with a historical abundance of food is the enormous complexity of the reasons for it. The people of this country and many other developed countries have come to expect abundant and cheap food. It has always been there, and there is no reason to expect anything different in the future. Most people base their expectations on experience and not on the many factors that make such an expectation a reality. It is flawed logic to assume that one will always have what one has historically had.

During my years as a weed science researcher, I was always amazed at the effectiveness of modern herbicides. Even those herbicides that were sometime inconsistent and performed erratically still impressed me. However, farmers always seemed to take for granted the efficacy of herbicides, which I saw as phenomenal. They often overlooked the complexity of the science involved in developing selective herbicides. Multiply this idea many times, and it becomes apparent that the average layperson has difficulty appreciating the enormous complexity of modern agriculture. Just what it takes in research to produce that extra bushel of wheat, the extra sweetness of a strawberry, or the super crispness of the Honey Crisp apple is not recognized or even appreciated. The many hours spent in the hot sun making countless crosses in wheat that produced the desired offspring that yielded that extra bushel are simply taken for granted.

Emergence of Competing Challenges

Another major reason for the loss of enthusiasm and support for agricultural research is that many experiments and investigations are under way at the present time that hold great interest for humankind. This involves every aspect of our existence, including areas such as medicine, defense, transportation, and the environment.

An old Chinese proverb says, "A person who has food has many problems. A person who has no food has only one problem." Obviously, agriculture's success has enabled many people to have many problems.

Success in medicine does two things: It solves a particular problem, and it whets the public's appetite for more successes or cures. Consequently, those successes have contributed to exponential growth in support for medical research to find a cure for all of life's threatening diseases. This simple fact has led to phenomenal budget increases in the National Institutes of Health in recent decades.

Another competing challenge is the dangerous and insecure world in which we live. In the twentieth century, the planet was seldom free of conflict. From the big wars, World Wars I and II, to a host of lesser (in scale) conflicts, including the Korean Conflict, Iraq, Afghanistan, and the Vietnam War, and skirmishes in Panama and Grenada. But the unleashing of nuclear power for destructive purposes in the twentieth century has added an additional dimension to people's justifiable concern. We must have research if we are to become more effective in killing or subduing the bad guys of the world.

Many of the federal government's expenditures center around individual health, safety, or national security. They fund much of the research in the Department of Defense, Homeland Security, Environmental Protection Agency, and Health and Human Services. My contention is that agriculture should be one of those important research areas supported by the federal government. While health for individuals and personal safety are important, the argument for a secure food supply is equally strong.

These are among the most important reasons that interest in and support for agricultural research has waned in recent decades. Recognizing each of these challenges will enable a better understanding of the problem of inadequate support for agricultural research. First, it is not possible to eliminate the emergence of competing challenges. A second challenge, the "historical abundance of plenty" is an emotional issue that cannot be dealt with by a scientific approach. Those who wish to strengthen agricultural research should focus on the "complexity of reasons accounting for an abundance of food." The average person on the street has little appreciation of what is required to ensure the abundance of food we enjoy. Of the four largest countries in the world, Russia, Canada, China, and the United States, we have the most arable land. There is no argument regarding the importance of arable land, but it is only a part of the picture. An enlightened regulatory system, available capital, effective communication and transportation system, knowledgeable farmers and information, as well as knowledge and technology, are all critical to successful agriculture. Production is only one part of the food system. What are the assumptions that must be met in order to ensure an ongoing abundant supply of relatively safe and cheap food?

Assumption 1: Availability of Human Capital

The entire supply chain, including producers (farmers), buyers, processors, manufacturers, storers, transporters, and so on, requires people who are willing to engage in these enterprises. In general, agriculture is the most successful enterprise in the developed world inasmuch as people have the freedom to choose their life's work. In this regard, choice is often driven by financial, social, or moral reward.

Is it logical to assume that societies, cultures, and governments worldwide will maintain a system of rewards that will cause future generations to choose to engage in commercial agriculture in a way that will allow a relatively small percentage of the population to provide nourishment for the remaining population? Will there be sufficient economic incentive? Will there be sufficient social incentive? Will government policy encourage this? Will government regulation discourage this? Then, assuming that sufficient numbers of people choose agriculture, what will be required for the next generation to be successful? The answer is knowledge, education, skill, expertise, and innovation. So, who will educate and train the next generation of agriculturists? The answer is the current generation of agriculturists, some of whom are agricultural researchers.

Feeding a growing world population will require proficiency and efficiency. Agricultural research drives innovation, which leads to proficiency and efficiency. As pointed out in chapter 9, much (perhaps most) of the recent success in agriculture can be accounted for by enhanced total factor productivity (TFP). Evidence for this is that output over the past half century has closely paralleled productivity. This has occurred while inputs have remained almost constant. Agricultural research expands the knowledge base, which is the key to education and enlightenment. It distinguishes truth from fiction and fact from conjecture. It makes it possible for an ever-shrinking producer population to feed an ever-expanding consumer population. It also demonstrates the potential for profit, which is essential in maintaining any enterprise.

Assumption 2: Availability of Energy

Agriculture is both a potential producer and a great user of energy. Agriculturists appreciate the fact that the sun is the ultimate source of energy and that plants are nature's most unique and most efficient means of converting solar energy into storable chemical energy. But they also know that modern, market-based, cash agricultural systems are very dependent on energy, especially fossil fuels (which, incidentally, exists because of plants). Whether one is talking about liquid petroleum fuels to power farm implements, natural gas needed to produce synthetic fertilizers, petroleum-based agricultural chemicals, plastics, rubber, or other products essential to agriculture, or whether one is talking about petroleum-based fuels that are required to power transportation systems that move agricultural products from producer to consumer, agriculture is highly dependent on fossil fuels.

Is it logical to assume that an abundant and affordable supply of energy will be available to "fuel" agriculture? An array of energy options will be available to the next generation of farmers. Is agriculture prepared, or will it be prepared, to adjust to the changes in global supply and demand, changes in global economic policy, and changes in geopolitical alliances that have such a profound effect on petroleum supply and price? Modest changes in energy costs could have major impacts on worldwide agricultural productivity. Agriculture is intrinsically involved in both food and energy security. Achieving both represents a major challenge for this planet.

Assumption 3: Availability of Water

Water is absolutely essential to agriculture. One could easily make the case that future worldwide agricultural production may be more dependent on water than on any other single resource. Water is the most limited of the major inputs to agricultural production—sunlight, nutrients, CO_2, and water. While much of the planet's arable land currently used for agricultural production typically receives adequate rainfall to support subsistence agriculture or modest harvestable animal and crop yields, the high-yield systems that will be required to feed a growing world population are not achievable without supplemental irrigation. Is it logical to assume that the current annual use of fresh surface water and/or groundwater for agricultural irrigation is sustainable? What about priorities? Who will establish them? How will agriculture fare in water wars where the competing interests of human/residential consumption, industry/commercial uses, recreation, agriculture, and others vie for a limited resource?

Again, agricultural research should play a pivotal role in developing and proving future agricultural systems that conserve precious fresh water while increasing agricultural outputs. Improved crop and animal genetics that lead to increased water-use efficiency, improved waste management and water recycling, and improved irrigation technology and management are just a few areas where agricultural research is essential.

Assumption 4: A Favorable Geopolitical Situation

Agriculture does not operate independently of politics. Public policy, or the absence thereof, has a huge impact on agriculture and agribusiness. For example, international trade, tariffs, subsidies, international monetary policy, and so on are all components of global politics. About one-third of US agricultural production is for export. The United States' relationships around the world are critical to the vitality of the nation's agriculture.

The importance of global politics is glaringly evident in the battle against hunger. It is often said that lack of effective distribution is the biggest problem in feeding the hungry people of the world. In fact, governmental instability, corruption, and war are far more common causes of hunger than is an inability to produce.

And, yes, agricultural research is essential to maintaining a favorable geopolitical situation. Agricultural research keeps American agriculture efficient, proficient, and competitive. Agricultural research in the United States has driven international agriculture, helping the world feed itself. A well-fed nation can become secure, stable, democratic, educated, and enlightened.

Though the focus is on the United States, these points are just as relevant to each country of the world. In fact, they are probably a greater concern for developing countries because these nations have less room for error.

As general support for agricultural research continues to decline in this country and in many others, it is evident that something must be done to reverse this trend. Clearly, it must be accepted that agricultural research has been the key that has established the potential of successful agriculture. Many definitive economic studies have shown the positive correlation between agriculture's success and agricultural research. Perhaps agriculture is a victim of its own success. Agricultural research is essential to improve efficiency and proficiency of production. It drives innovation and change. But it also can be a critical means of risk management relative to possible changes in the big four—human capital, energy, water, and politics. Research will ensure education, training, and motivation of the next generation of agriculturists. Research will provide solutions to energy and water challenges, and it will enable agriculture to adapt. Finally, agriculture and world politics are mutually dependent on one another. Peace without adequate food is unattainable.

Chapter 11 outlines the expectations and responsibilities of all who benefit from agricultural research. Clearly since all benefit, all should have a role to play in supporting this system. This leads to the pertinent question, who can and will lead and participate in this effort? Who has the interest? The answer may be another question: Who has the most to lose if the meeting is not called? The answer is everyone. But who stands to gain? Again, the answer is everyone. The real question is, who has the clout to call the meeting, and who is willing to invest their clout, time, and effort in doing so? Though it would be nice to have a global conference on agricultural research, let us not set too lofty a goal. Can we get a conference in the United States? Who will call for the meeting? Will it be the president? Probably not. While presidents Washington, Jefferson, and Lincoln were highly knowledgeable about agriculture, few, if any, of the presidents in the past 100 years have had any real knowledge or visible appreciation of agriculture, especially agricultural research.[5]

Maybe the secretary of agriculture could take a forceful lead in supporting agricultural research. Unfortunately, the secretary of agriculture has many other politically more sensitive concerns (e.g., nutrition programs to feed the food-insecure population, conservation programs for the landowner, farm programs to support certain producers, regulatory programs). Consequently, little time is left for agricultural research. Could agribusiness giants take the lead? Hardly. Such organizations have already taken the lead in agricultural research that they pay for and that benefits their own bottom line. This is about all we can reasonably expect from corporate America.

Commodity and trade associations could conceivably take the lead, but my experience is that such groups almost always focus on their specific concerns rather than the overall common good. Every group mentioned in chapter 11 that has some power base probably cannot be expected to be a leading advocate for agricultural research. One hope is academia. Unfortunately, this group has essentially a zero power base. Perhaps leadership could come from banking, finance, and trade groups coupled with academia. This would be a new model, but since the old models are not working, it is worth a try. While I am not providing a comforting solution to identifying leadership for the challenge, all groups should rethink their priorities. Realistically, meeting the agricultural research challenge will require a coordinated effort by all who benefit from agriculture. Ergo, the obvious complexity.

Over the past 100 years a shared responsibility for agricultural research resulted in a period of great accomplishment. On the public front, state and federal governments provided significant financial support for agricultural research. This country's land-grant university system, the USDA, state departments of agriculture, and other public entities worked together to provide leadership in agricultural research. The luster and appeal of agricultural research seems to have waned for both the USDA and the land-grant system as the USDA has embraced a myriad of programs in nutrition, rural development, housing, social programs, and so on.

As land-grant institutions become larger, such growth is usually in areas other than agriculture. While agriculture and the mechanical arts (engineering) were the key components of the original concept of the land-grant university, today they are being marginalized to a great extent. In fact, in some institutions colleges of agriculture have morphed into a college of agriculture and food sciences; college of agriculture, life, and natural sciences; college of agriculture, home economics, and allied programs; school of natural resources and agricultural sciences; college of agriculture, environment, and nutrition sciences; college of agriculture and technology; college of tropical agriculture and human resources; college of agriculture, biotechnology, and natural resources; and college of agricultural and environmental sciences. You will note that in some institutions, agriculture does not even get top billing in the name of the college that includes the agricultural sciences.

Concern about the future of agriculture and agricultural research requires a deeper commitment to education in the agricultural sciences. Who will train the next generation of agricultural workers? Who will train the next generation of agricultural scientists? Even though the land-grant system has historically invested heavily in graduate education, thus producing competence in research, a concern exists that certain areas of science (disciplines) are being lost. For example, the composition of student bodies in agricultural colleges across the country illustrates some common trends: substantial growth in the number of students in agricultural economics and business, growth in biosystems engineering, and growth in animal science (often as a precursor to medical or veterinary school). However, there are sometimes troubling declines in agronomy, soil science, entomology, plant physiology, plant pathology, plant breeding, and agricultural engineering. Investment in a broad range of agricultural research will not only produce new knowledge, innovation, efficiency, and so on but also contribute to training the next generation of agriculturists. This is essential.

Agriculture made possible our civilization as we know and enjoy it today. More importantly, our future is totally contingent upon the success of agriculture in the years to come. For this to occur, the agricultural enterprise must be supported by a robust research effort. Equally important is the fact that everyone has a responsibility to carry out and a role to play. For the sake of our well-being, we must make a concerted effort to strengthen the support for agricultural research. In order to achieve success, we must move forward! In this final chapter I have considered the future and how we will get there. My final assessment is that the key is leadership. With a challenge so great, one hopes for the emergence of one or more leaders who will be able to articulate the role of agricultural research such that its importance and relevance are universally recognized and appreciated. With that accomplished, our future becomes a bit more secure.

We must be successful. Indeed, we have no other choice.

Notes

Chapter 2

1. The idea of using what you have to solve a problem was a practical approach. Using some of the nation's abundance of public lands to support public education was both a practical and a powerful approach. In a speech (the York Lecture) at Auburn University in 2012 I proposed that perhaps it was time for the United States to consider another land sale, with the proceeds going to support agricultural research. The nation still owns millions of acres of land, particularly in the western part of the country. Many people believe that, with the exception of land for roads, military reservations, nature reserves, parks and recreational facilities, and other common land-use areas, that land could be more effectively managed by the private sector.

2. Another three-legged-stool concept embraces the primary groups engaged in agricultural research in the United States. These include the State Agricultural Experiment Stations, the USDA and other federal agencies, and industry. Each is a major contributor to the success of this enterprise.

3. Provision for a research mission focusing on the 1890 institutions was made in the 1977 Farm Bill, Sec. 1445 (7 U.S.C. 3222), on agricultural research, at 1890 land-grant colleges, including Tuskegee University.

4. Legislation titled "Improving America's Schools Act of 1994 (H.R.6), section 531 (Equity in Educational Land-Grant Status Act), provided for 1994 land grant institutions. It became effective on October 20, 1994, as Public Law no. 103–382.

Chapter 3

1. On a personal note, I once was told that the local government had gone so far as to survey and get an appraisal for a tract of experiment station land before I learned of the proposed land grab! Fortunately, these plans by the local government did not work out, so we were able to keep our land.

Still another potential land grab occurred within the academic family. One day when returning from a meeting, I observed a large group out in one of our scientist's plant nurseries. After a few inquiries I learned they were representatives from the local trade school. At the state level they had asked the University System Board of

Regents for some help in obtaining land. Someone at the Board of Regents apparently told them that the Experiment Station had thousands of acres of land—which was true. The group then proceeded to select a 40-acre tract of prime research land with ready access to major highways. They had done their homework with the university president and the University System Board of Regents. However, they had not gone to the legislature. So I made my case to the House and Senate Agriculture Committees and a few other influential members of the state legislature. I was fortunate in that members of the legislature trumped university officials and the Board of Regents. As a result, the land remained with the experiment station.

2. Seaman A. Knapp, physician, college instructor, and farmer played a key role in the development of legislation that provided the basis for the Hatch Act of 1887, which laid the foundation for a nationwide network of agricultural experiment stations. An archway over Independence Avenue in Washington, DC, connecting the USDA headquarters (North building)—the Whitten Building—with the USDA South building is named in honor of Seaman Knapp.

3. The Hatch Act of 1887, which provided for the state Agricultural Experiment Stations, should not be confused with the Hatch Act of 1939. The 1939 Hatch Act was aimed at preventing improper political practices and prevented civil servants from campaigning.

4. To develop a funding base for a College of Agriculture, the North Carolina General Assembly enacted a law imposing a tax of one dollar on every dog in the state for the benefit of the state treasury and the College of Agriculture. It was reported the resolution was passed with great unanimity.

5. Fifteen thousand dollars in 1887 would be equivalent to $375,000 in 2014.

6. This organization has undergone a number of name changes, including Association of Land-Grant Colleges in 1920; Association of Land-Grant Colleges and Universities in 1926; American Association of Land-Grant Colleges and State Universities in 1955; Association of State Universities and Land-Grant Colleges in 1962; and National Association of State Universities and Land-Grant Colleges (NASULGC) in 1965. The latest name change came in 2009, when the organization was renamed the Association of Public and Land-Grant Universities (APLU).

7. The Board on Agriculture Assembly is one of the boards that make up the Association of Public and Land-Grant Universities. The Board on Agriculture Assembly was previously called simply Board on Agriculture.

8. The committee of nine included two members from each of the four regional experiment station director associations and one member representing the Home Economics Association.

Chapter 4

1. Hybrid corn emerged from a unique source of germplasm developed by George Sprague, a USDA scientist at Iowa State University. Sprague was a strong proponent of recurrent selection, a constant recycling of the cream of the crop from each breeding cycle. The result of this effort was the evolution of elite inbred lines well suited for hybrid development. These lines, generally referred to as "B" lines, delivered a solid package of valued traits, including yield, root and stalk structure, disease and insect resistance, and maturity. More important, they performed exceptionally well when crossed with other lines. The B lines were used to create the hybrid varieties of corn that are grown on millions of acres of land throughout the United States and around the world. Much of the increase in yield of corn is the result of genetic improvement, much of which can be linked to the B lines.

Chapter 5

1. Data for each group category include varieties with both herbicide tolerant (HT) and Bt (*Bacillus thuringiensis*) stacked traits.

Chapter 7

1. An unexpected side effect of the controlled atmosphere of apples was the loss of several marginal apple-growing regions of Appalachia.

Chapter 8

1. Today, more and more public institutions are seeking to capitalize on intellectual property. As they try to benefit from such discoveries, they become more like industry.

2. Dr. Keith Fuglie, USDA/ERS, personal communication, January 2013.

3. I readily acknowledge that aquaculture and mariculture provide a portion of the food that sustains us and that there are perhaps some means of using the world's waters as an energy source.

Chapter 9

1. Presentation at the Borlaug Dialogue International Symposium, October 2014.

2. Oxisols are an order in soil taxonomy occurring primarily in rain forests. They have a low cation exchange capacity with no more than 10% weatherable minerals. These soils are red or yellowish in color and may contain iron, aluminum oxides, hydroxides, and quartz, along with small amounts of other minerals and organic matter.

3. Pyrolysis is the thermochemical decomposition of organic material that takes place at high temperatures and in the absence of oxygen. By-products include heat, syngas (a mixture of hydrogen and carbon monoxide), and biochar. Pyrolysis has application, for example, in producing syngas and biochar from sawdust or food waste.

4. A "once-through" basis means that water is not recycled.

Chapter 10

1. Appropriations data for NIFA include research and education. No attempt is made to separate the research component from educational functions in the NIFA budget, although research constituted the majority of that budget. Because of the difficulty of separating the budget for these functions, I have chosen to refer to the total NIFA budget (excluding Extension) as agricultural research.

2. Oftentimes it is a committee of key scientists or scientific leaders from across the country.

Chapter 11

1. This position often goes by other names (e.g., department chair, department chairperson).

2. Hearings were held in each state with the exception of Louisiana.

3. The Tri-Societies comprise the American Society of Agronomy, the Crop Science Society of America, and the Soil Science Society of America.

Chapter 12

1. This development gave new life to the herbicide glyphosate, which has been one of the most successful agrochemicals in history. In response to concerns about the safety of glyphosate use, a number of studies are currently assessing its health risks.

Chapter 13

1. Inflation-adjusted purchasing power parity (ppp) = 2005 ppp dollars.

Chapter 14

1. I make no effort to indicate the carrying capacity of the planet, but as all ranchers know, such is the nature of land use. For long-term sustainability, I suspect that we are approaching (if we have not already surpassed) the sustainable carrying capacity of the planet while maintaining a desirable environment and appropriate biodiversity at the current levels of technology.

2. Until this provision was made, the University of Georgia had not had a College of Engineering since it accepted the conditions of the Land Grant Bill in 1872. Acceptance of the provisions of the Morrill Act required a commitment to both agriculture *and* the mechanical arts.

3. I acknowledge that this recommendation would supersede some others.

4. I greatly acknowledge the contribution of Dr. David Bridges to this section. Earlier in his career Dr. Bridges was a distinguished agricultural researcher. Today he is president of the Abraham Baldwin Agricultural College in Tifton, Georgia.

5. I acknowledge that President Jimmy Carter had some working, practical knowledge of agriculture.

Bibliography

Introduction

Cho, Adrian. "The Discovery of the Higgs Boson." *Science* 338 (2012): 1524–25.

———. "Higgs Boson Makes Its Debut after Decades-Long Search." *Science* 337 (2012): 141–42.

Grant, Edward. "When Did Modern Science Begin?" *American Scholar* 66 (1997): 105–14.

Molina, Mario, Ernest J. Moniz, Craig Mundie, Ed Penhoet, Barbara Schaal, Eric Schmidt, Daniel Schrag, et al. *Report to the President on Agricultural Preparedness and the Agriculture Research Enterprise* (2012). President's Council of Advisors on Science and Technology. http://www.whitehouse.gov/sites/default/files/microsites/ostp/pcast_agriculture_20121207.pdf

Population Reference Bureau. *World Population Data Sheet* (2009). Retrieved May 8, 2015, from http://www.prb.org/pdf09/09wpds_eng.pdf

Population Reference Bureau. *World Population Data Sheet* (2012). Retrieved May 8, 2015, from http://www.prb.org/pdf12/2012-population-data-sheet_eng.pdf

Chapter 1

Borlaug, Norman E. *Agricultural Science and the Public* (1973). Council for Agricultural Science and Technology, paper no. 1.

China's Motor Vehicles Top 233 Mln. Retrieved June 18, 2015, from http://news.xinhuanet.com/english/china/2012–07/17/c_131721176.htm

Countries of the World. Worldatlas. Retrieved February 10, 2014, from http://www.worldatlas.com/aatlas/populations/ctypopls.htm#.UvkQpPldVHU

Graham, L. R. *Science in Russia and the Soviet Union.* New York: Cambridge University Press, 1993.

Human Population Dynamics. Section 4: World Population Growth through History. Retrieved April 4, 2014, from http://www.learner.org/courses/envsci/unit/text.php?unit=5&secNum=4

India General Data. 2012, November 29. http://indiatransportportal.com/2012/11/vehicles-in-india/

Pew Research Center. 2014, February 3. Retrieved February 10, 2014, from http://www.pewresearch.org/fact-tank/2014/02/03/10-projections-for-the-global-population-in-2050/

Solutions from the Land. 2013. www.SFLdialogue.net

Total Population of the World by Decades, 1950–2050. Retrieved February 10, 2014, from http://www.infoplease.com/ipa/A0762181.html

U.S. Census Bureau World Population Clock. Retrieved April 4, 2014, from http://census.gov/popclock

World Factbook. http://www.cia.gov/library/publications/the-world-factbook/index.html

World Bank Data. Retrieved April 4, 2014, from http://data.worldbank.org/data-catalog/world-development-indicators

Chapter 2

Edmond, J. B. *The Magnificent Charter: The Origin and Role of the Morrill Land-Grant Colleges and Universities.* Hicksville, NY: Exposition Press, 1978.

First Morrill Act. Passed by the US Congress and signed into law by President Abraham Lincoln on July 2, 1862.

Wade, Kathryn Lindsay Anderson. "The Intent and Fulfillment of the Morrill Act of 1862: A Review of the History of Auburn University and the University of Georgia." Master's thesis, Auburn University, 2005.

Chapter 3

Bishop, W. D. *Historical Sketch of the United States Agricultural Society.* U.S. Patent Office Report (pt. 2, Agr.) (1860), 22–30.

Carpenter, William L., and Dean W. Colvard. *Knowledge Is Power.* Raleigh: North Carolina State University, 1987.

Carrier, Lyman. *The Beginnings of Agriculture in America.* New York: McGraw-Hill, 1923.

Edmond, J. B. *The Magnificent Charter: The Origin and Role of the Morrill Land-Grant Colleges and Universities.* Hicksville, NY: Exposition Press, 1978.

Huffman, Wallace E., and Robert E. Evenson. *Sciences for Agriculture.* Ames: Iowa State University Press, 1993.

Jordan, John Patrick, Gale A. Buchanan, Neville P. Clarke, and Kelly C. Jordan. *Leadership in Agriculture: Case Studies for a New Generation.* College Station: Texas A&M Press, 2013.

Kerr, N. A. *A History of the Alabama Agricultural Experiment Station, 1883–1983.* Auburn, AL: Agricultural Experiment Station, 1985.

———. *The Legacy: A Continual History of the State Agricultural Experiment Stations, 1887–1987.* Columbia: Missouri Agricultural Experiment Station, University of Missouri, 1987.

Rothamsted Experimental Station. Lawes Agricultural Trust. 1977.

True, Alfred C. *A History of Agricultural Experimentation and Research in the United States: 1607–1925.* USDA Miscellaneous Publication 251. Washington, DC: USGPO, 1937.

von Liebig, Justus. *Organic Chemistry in Its Applications to Agriculture and Physiology.* Edited from the Manuscript of the Author by Lyon Playfair, Ph.D. Printed for Taylor & Walton, Booksellers and Publishers to University College, Upper Gower Street, London (1840).

Wade, Kathryn Lindsey Anderson. "The Intent and Fulfillment of the Morrill Act of 1862: A Review of the History of Auburn University and the University of Georgia." Master's thesis, Auburn University, 2005.

Woodard, Carl Raymond, and Ingrid Nelson Waller. *New Jersey's Agricultural Experiment Station, 1880–1930.* New Brunswick: New Jersey Agricultural Experiment Station, 1932.

Chapter 4

Bradley, Daniel G. "Genetic Hoof Prints." *Natural History Magazine* 112(1) (2003): 35–41.

———, Ronan T. Loftus, Patrick Cunningham, and Dave E. MacHugh. "Genetics and Domestic Cattle Origins." *Evolutionary Anthropology* 6 (1998): 79–86.

Diamond, Jared. *Guns, Germs, and Steel: The Fates of Human Societies.* New York: Norton, 1999.

Collings, Gilbeart H. *Commercial Fertilizers: Their Sources and Uses.* New York: McGraw Hill Book Companion, 1955.

Götherström, Anders, Cecilia Anderung, Linda Hellborg, Rengert Elburg, Colin Smith, Dan G. Bradley, and Hans Ellegren. "Cattle Domestication in the Near East Was Followed by Hybridization with Aurochs Bulls in Europe." *Proceedings of the Royal Society of Biological Sciences* 272(1579) (2005): 2345–50. doi: 10.1098/rspb.2005.3243

A History of American Agriculture, 1607–2000. ERS POST 12. Washington, DC: USDA, Economic Research Service, 2000.

The History of the Automobile: The Internal Combustion Engine and Early Gas-Powered Cars, retrieved September 11, 2009, from http://inventors.about.com/library/weekly/aacarsgasa.htm?rd=1; retrieved October 19, 2009, from http://inventors.about.com/library/weekly/aacarsgasa.htm?rd=1

Lawes, J. B., J. H. Gilbert, and Evan Pugh. "On the Sources of the Nitrogen of Vegetation; with Special Reference to the Question Whether Plants Assimilate Free or Uncombined Nitrogen." Abstract, *Proceedings of the Royal Society of London* 10 (1860): 544–57.

Hall, K., and E. Nowels. 2010. *Fertilizer 101*. Washington, DC: Fertilizer Institute.

Lewis, W. J., and J. H. Tumlinson. "Host Detection by Chemically Mediated Associative Learning in a Parasitic Wasp." *Nature* 331 (1988): 257–59.

Miles, Randall J., and James R. Brown. "The Sanborn Field Experiment: Implications for Long-Term Soil Organic Carbon Levels." *Agronomy Journal* 103 (2011): 268–78.

Mullen, R. W., K. W. Freeman, G. Johnson, and W. R. Raun. "The Magruder Plots: Long-Term Wheat Fertility Research." *Better Crops* 85 (2001): 6–8.

Rains, G. C., S. L. Utley, and W. Joe Lewis. "Behavioral Monitoring of Trained Insects for Chemical Detection." *Biotechnology Progress* 22 (2006): 2–8.

Rothamsted Experimental Station. Lawes Agricultural Trust. 1977.

Sinclair, T. R., and C. J. Sinclair. *Bread, Beer and Seeds of Change: Agriculture's Impact on World History*. Wallingford, Oxfordshire, UK: CABI, 2010.

Sisler, Harry H., Calvin A. Vander Werf, and Arthur W. Davidson. *College Chemistry: A Systematic Approach*. New York: Macmillan, 1953.

Slater, Joseph V., and Bill J. Kirby. *Commercial Fertilizers 2010*. Columbia, MO: Association of American Plant Food Control Officials, Fertilizer/Ag Lime Control Service, 2011.

United Nations, Population Division. 2009, October 1. *World Population Prospects: The 2006 Revision and World Urbanization Prospects* (New York: United Nations, Department of Economic and Social Affairs, 2008). http://esa.un.org/unpp

von Liebig, Justus. *Organic Chemistry in Its Applications to Agriculture and Physiology*. Edited from the Manuscript of the Author by Lyon Playfair, Ph.D. Printed for Taylor & Walton, Booksellers and Publishers to University College, Upper Gower Street, London (1840).

Chapter 5

"The End of the Oil Age." 2003. *Economist* 369: 11–12. Retrieved June 3, 2015, from http://www.economist.com/node/2155717

Fernandez-Cornejo, Jorge. "Adoption of Genetically Engineered Crops in the U.S." Data Products Topic Page. Washington, DC: USDA, Economic Research Service, 2008, September 1. Retrieved September 10, 2013, from http://www.ers .usda.gov/data-products/adoption-of-genetically-engineered-crops-in-the-us/ recent-trends-in-ge-adoption.aspx

———. "Rapid Growth in Adoption of Genetically Engineered Crops Continues in U.S." *Amber Waves* (2008).

———, and Margriet Caswell. *The First Decade of Genetically Engineered Crops in the United States.* Economic Information Bulletin no. 11. Washington, DC: USDA, Economic Research Service, 2006.

Hess, Charles E., Randolph Barker, Lawrence Bogorad, Ralph E. Christoffersen, Albert H. Ellingboe, Anthony Faras, Jack Gorski, et al. *Agricultural Biotechnology: Strategies for National Competitiveness.* National Research Council, National Academy of Science. Washington, DC: National Academies Press, 1987.

Laughlin, Robert B. Powering the Future: How We Will (Eventually) Solve the Energy Crisis and Fuel the Civilization of Tomorrow. New York: Basic Books, 2011.

Martenson, Chris. *The Crash Course.* Hoboken, NJ: Wiley, 2011.

Mervis, Jeffrey. "The Battle over the 2011 Budget: What's at Stake for Research." *Science* 331 (2011): 14–15.

National Science Foundation. 2014. *Basic Research to Enable Agricultural Development (BREAD).* Retrieved June 8, 2015, from https:www.nsf.gov/funding/ pgm_summ.jsp?pims_id=503285

The Nobel Prize in Physiology or Medicine 1948. NobelPrize.org. Retrieved July 26, 2007, from http://nobelprize.org/nobel_prizes/medicine/laureates/1948/

Sinclair, T. R., and C. J. Sinclair. *Bread, Beer and the Seeds of Change: Agriculture's Impact on World History.* Wallingford, Oxfordshire, UK: CABI, 2010.

Specter, Michael. "Seeds of Doubt." *New Yorker* (2014, August 25), 46–57.

Watson, J. D., and F. H. C. Crick. "A Structure for Deoxyribose Nucleic Acid." *Nature* 171 (1953): 737–38.

Xu, Kenong, Xu Xia, Fukao Takeshi, Patrick Canlas, Reycel Maghirang-Rodriguez, Sigrid Heuer, Adbelbagi M. Ismail, et al. "*Sub 1 A* is an Ethylene-Response Factor-Like Gene That Confers Submergence Tolerance to Rice." *Nature* 442 (2006): 705–708.

Chapter 6

Black, Robert E., Lindsay H. Allen, Z. A. Bhutta, Laura E. Caulfield, Mercedes de
   Onis, Majid Ezzati, Colin Mathers, et al. "Maternal and Child Undernutrition:
   Global and Regional Exposures and Health Consequences." *Lancet* 371 (2008):
   243–60.

Brazzel, J. L., and L. D. Newsom. 1959. "Diapause in *Anthonomus grandis* Boh."
   Journal of Economic Entomology 52: 603–11.

Paine, Jacqueline A., Catherine A. Shipton, Sunandha Chaggar, Rhian M. Howells,
   Mike J. Kennedy, Gareth Vernon, Susan Y. Wright, et al. "Improving the Nutri-
   tional Value of Golden Rice through Increased Pro-Vitamin A Content." *Nature
   Biotechnology* 23 (2005): 482–87.

Römer, Susanne, Paul D. Fraser, Joy W. Kiano, Cathie A. Shipton, Norihiko Misawa,
   Wolfgang Schuch, and Peter M. Bramley. "Elevation of the Provitamin A Content of Transgenic Tomato Plants." *Nature Biotechnology* 18 (2000): 666–69.

Ye, X., S. Al-Babili, A. Klöti, J. Zhang, P. Lucca, P. Beyer, and I. Potrykus. "Engineering the Provitamin A (Beta-Carotene) Biosynthetic Pathway into (Carotenoid-
   Free) Rice Endosperm" *Science* 287(5451) (2000): 303–305.

Chapter 7

Coyne, George. 2009. "A Tragic Mutual Incomprehension." Retrieved June 8, 2015,
   from http://www.resetdoc.org/story/00000001505

Food and Agriculture Organization of the United States. "The State of Food Insecurity in the World" (2011). Retrieved May 7, 2015, from http://www.fao.org/
   docrep/014/i2330e/i2330e00.htm

Huffman, W. E. "Household Production and the Demand for Food and Other Inputs: U.S. Evidence." *Journal of Agricultural and Resource Economics* 36 (2011):
   465–87.

Chapter 8

Alston, Julian M., Matthew A. Andersen, Jennifer S. James, and Philip G. Pardey.
   *Persistence Pays: U.S. Agricultural Productivity Growth and the Benefits from
   Public R&D Spending.* New York: Springer, 2010.

Alston, Julian M., Matthew A. Andersen, Jennifer S. James, and Philip G. Pardey.
   "The Economic Returns to U.S. Public Agricultural Research." *American Journal
   of Agricultural Economics* 93 (2011): 1257–77.

Charles, Daniel. *Lords of the Harvest: Biotech, Big Money, and the Future of Food.* Cambridge, MA: Perseus, 2001.

Economic Research Service. *Agricultural Productivity in the U.S.* Washington, DC: USDA. Retrieved May 7, 2015, from http://www.ers.usda.gov/data-products/agricultural-productivity-in-the-us.aspx (last updated June 13, 2014).

Fuglie, Keith O. "Total Factor Productivity in the Global Agricultural Economy: Evidence from FAO Data." In J. M. Alston, B. A. Babcock, and P. G. Pardey (Eds.), *The Shifting Patterns of Agricultural Production and Productivity World Wide* (pp. 63–95). Ames: Midwest Agribusiness Trade Research and Information Center, Iowa State University, 2010.

——, N. Ballenger, K. Day, C. Klotz, M. Ollinger, J. Reilly, U. Vasavada, et al. "Agricultural Research and Development: Public and Private Investments under Alternative Markets and Institutions." Agricultural Economic Report no. 735. Washington, DC: USDA, Economic Research Service, 1996, May.

Fuglie, Keith O., and P. W. Heisey. "Economic Returns to Public Agricultural Research." Economic Brief no. 10. Washington, DC: USDA, Economic Research Service, 2007, September.

Fuglie, Keith O., J. M. MacDonald, and Eldon Ball. "Productivity Growth in U.S. Agriculture." Economic Brief no. 9. Washington, DC: USDA, Economic Research Service, 2007, September.

Hauser, E. W., and G. A. Buchanan. "Influence of Row Spacing, Seeding Rates and Herbicide Systems on the Competitiveness and Yield of Peanuts." *Peanut Science* 8 (1981): 74–81.

Heisey, Paul, Sun Ling Wang, and Keith O. Fuglie. "Public Agricultural Research Spending and Future U.S. Agricultural Productivity Growth: Scenarios for 2010–2050." Economic Brief no. 17. Washington, DC: USDA, Economic Research Service, 2011, July.

Huffman, W. E., and R. E. Evenson. "Do Formula or Competitive Grant Funds Have Greater Impact on State Agricultural Productivity?" *American Journal of Agricultural Economics* 88 (2006): 783–98.

——. *Science for Agriculture: A Long-Term Perspective* (2nd ed.). Ames, IA: Blackwell, 2006.

Huffman, W. E., G. Norton, and L. G. Tweeten. "Investing in a Better Future through Public Agricultural Research." CAST Commentary (TA 2011–1, 2011, March).

Jin, Y. and Huffman, W. E., "Measuring Public Agricultural Research and Extension and Estimating their Impacts on Agricultural Productivity: New Insights from US Evidence." *Agricultural Economics* 46 (2015): forthcoming.

Plastina, Alejandro, and Lilyan Fulginiti. "Rates of Return to Public Agricultural Research in 48 US States." *Journal of Productivity Analysis* 37 (2012): 95–113.

Wang, S. L., V. E. Ball, L. E. Fulginiti, and A. Plastina. "Accounting for the Impact of Local and Spill-In Public Research, Extension and Roads on U.S. Regional Agricultural Productivity, 1980–2004." In K. Fuglie, S. L. Wang, and V. E. Ball (Eds.), *Productivity Growth in Agriculture: An International Perspective* (pp. 13–32). Wallingford, UK: CAB International, 2012.

Chapter 9

Barber, Nancy L. *Summary of Estimated Water Use in the United States.* Fact Sheet 2009–3098. U.S. Department of the Interior, U.S. Geological Survey (2009). Retrieved May 7, 2015, from Pubs.usgs.gov/fs/2009/3098/

Buchanan, Gale A., Robert W. Herdt, and Luther G. Tweeten, *Agricultural Productivity Strategies for the Future: Addressing U.S. and Global Challenges.* Issue Paper no. 45 (2010), Council for Agricultural Science and Technology, Ames, IA.

Canning, Patrick, Charles Ainsley, Sonya Huang, Karen R. Polenske, and Arnold Waters. "Energy Use in the U.S. Food System." USDA Economic Research Service Report no. ERR-94 (2010). Retrieved May 7, 2015, from http://www.ers.usda.gov/publications/err-economic-research-report/err94.aspx

Elwell, Frank W. *A Commentary on Malthus' 1798 Essay on Population as Social Theory.* Lewiston, NY: Mellen Press, 2001.

———. *Malthus' Social Theory* (2001). Retrieved May 7, 2015, from http://www.faculty.rsu.edu/users/f/felwell/www/Theorists/Malthus/SocMalthus.htm

———. "Reclaiming Malthus" (2001). Retrieved December 22, 2012, from http://www.faculty.rsu.edu/~felwell/Theorists/Malthus/reclaim.htm

Gunders, Dana. "Wasted: How America Is Losing up to 40 Percent of Its Food from Farm to Fork to Landfill." Natural Resources Defense Council Issue Paper (2012, August).

Hatch, M. D., and C. R. Slack. "Photosynthesis by Sugar Cane Leaves." *Biochemical Journal* 101 (1966): 103–11.

Lowdermilk, Walter Clay. "Conquest of the Land through Seven Thousand Years." USDA Bulletin no. 99. National Resources Conservation Service (1953).

Li, Shizhong, Guangming Li, Lei Zhang, Zhixing Zhou, Bing Han, Wenhui Hou, Jingbing Wang, et al. "A Demonstration Study of Ethanol Production from Sweet Sorghum Stems with Advanced Solid State Fermentation Technology." *Applied Energy* 102 (2013): 260–65.

Malthus, T. R. "An Essay on the Principle of Population" (1798). Retrieved December 22, 2012, from http://www.gutenberg.org/files/4239/4239-h/4239-h.htm

Oerke, E.-C., and H.-W. Dehne. "Safeguarding Production: Losses in Major Crops and the Role of Crop Protection." *Crop Protection* 23 (2004): 275–85.

Pettis, Jeffrey, and Keith S. Delaplane. "Coordinated Responses to Honey Bee Decline in the USA." *Apidologie* 41(3) (2010): 256–63.

Saintenac, Cyrille, Wenjun Zhang, Andres Salcedo, Matthew N. Rouse, Harold N. Trick, Eduard Akhunov, and Jorge Dubcovsky. "Identification of Wheat Gene Sr35 That Confers Resistance to Ug99 Stem Rust Race Group." *Science* 341 (2013): 783–86.

Schnepf, Randy. "Energy Use in Agriculture: Background and Issues." CRS Report for Congress (2004). Retrieved December 22, 2012, from http://nationalaglaw center.org/wp-content/uploads/assets/crs/RL32677.pdf

Slack, C. R., and M. D. Hatch. "Comparative Studies on the Activity of Carboxylases and Other Enzymes in Relation to the New Pathway of Photosynthetic Carbon Dioxide Fixation in Tropical Grasses." *Biochemical Journal* 103(3) (1967): 660–65.

Thompson, Robert L. "Proving Malthus Wrong: Sustainable Agriculture in 2050." *Tomorrow's Table* (2011). Retrieved May 7, 2015, from http://scienceblogs.com/ tomorrowstable/2011/05/13/proving-malthus-wrong-sustaina/

United Nations, Population Division. *World Population Prospects: The 2006 Revision and World Urbanization Prospects* (2008). Department of Economic and Social Affairs, New York. Retrieved October 1, 2009, from http://esa.un.org/ unpp

Normile, Dennis. "Driven to Extinction." *Science* (319) (2008): 1606–1609.

Woteki, Catherine. "The Road to Pollinator Health." *Science* 341 (2013, August 16): 695.

Chapter 10

Food, Conservation, and Energy Act of 2008. Public Law 110–234 (Farm Bill).

Goodwin, B. K., and Vincent H. Smith. "Theme Overview: The 2014 Farm Bill: An Economic Welfare Disaster or Triumph?" *Choices* (2014, 3rd quarter). Retrieved May 7, 2015, from http://www.choicesmagazine.org/choices-magazine/theme-articles/3rd-quarter-2014/theme-overview-the-2014-farm-billan-economic-welfare-disaster-or-triumph

Heisey, Paul, Sun Ling Wang, and Keith Fuglie. "Public Agricultural Research

Spending and Future U.S. Agricultural Productivity Growth: Scenarios for 2010–2050." Economic Brief no. 17. Washington, DC: USDA, Economic Research Service, 2011, July.

Huffman, W. E. "Measuring Public Agricultural Research Capital and Its Impact on State Agricultural Productivity in the U.S." Mimeograph (2010), Iowa State University, Department of Economics, Ames, IA.

———, and R. E. Evenson. "Do Formula or Competitive Grant Funds Have Greater Impact on State Agricultural Productivity? *American Journal of Agricultural Economics* 88 (2006): 783–98.

Huffman, W. E., and R. E. Just. "Setting Efficient Incentives for Agricultural Research: Lessons from Principal-Agent Theory." *American Journal of Agricultural Economics* 82 (2000): 828–41.

Kerr, Norwood Allen. *The Legacy: A Centennial History of the State Experiment Stations, 1887–1987*. Columbia: Missouri Agricultural Experiment Station, University of Missouri–Columbia, 198.

Molina, Mario, Ernest J. Moniz, Craig Mundie, Ed Penhoet, Barbara Schaal, Eric Schmidt, Daniel Schrag, et al. *Report to the President on Agricultural Preparedness and the Agriculture Research Enterprise* (2012). President's Council of Advisors on Science and Technology. Retrieved May 7, 2015, from http://www.whitehouse.gov/sites/default/files/microsites/ostp/pcast_agriculture_20121207.pdf

Office of Technology Assessment. *An Assessment of the United States Food and Agriculture Research System*. Library of Congress Catalog no. 81600189. Washington, DC: USGPO, 1981.

Rothamsted Experimental Station. Lawes Agricultural Trust. 1977.

Tegene, Abebayehu, Anne Effland, Nicole Ballenger, George Norton, Albert Essel, Gerald Larson, and Winfrey Clarke. *Investing in People: Assessing the Economic Benefits of 1890 Institutions*. Miscellaneous Publication no. 1583. Washington, DC: USDA, Economic Research Service, 2002.

von Liebig, Justus. *Organic Chemistry in Its Applications to Agriculture and Physiology*. Edited from the Manuscript of the Author by Lyon Playfair, Ph.D. Printed for Taylor & Walton, Booksellers and Publishers to University College, Upper Gower Street, London (1840).

Chapter 11

Buchanan, Gale A., Robert W. Herdt, and Luther G. Tweeten. *Agricultural Productivity Strategies for the Future: Addressing U.S. and Global Challenges*. Issue Paper

45 (2010), Council for Agricultural Science and Technology, Ames, IA.

Herdt, Robert W. "People, Institutions, and Technology: A Personal View of the Role of Foundations in International Agricultural Research and Development, 1960–2010." *Food Policy* 37 (2012): 179–90.

Huffman, W. E., G. Norton, and L. G. Tweeten. "Investing in a Better Future through Public Agricultural Research." CAST Commentary (TA 2011–1, 2011, March).

Jordan, John Patrick, Gale A. Buchanan, Neville P. Clarke, and Kelly C. Jordan. *Leadership in Agriculture: Case Studies for a New Generation.* College Station: Texas A&M Press, 2013.

Mervis, Jeffrey. 2014, January 17. "Final 2014 Budget Helps Science Agencies Rebound." *Science* 343(6168): 237. doi: 10.1126/science.343.6168.237

———. 2012, January 6. "Research Remains a Favored Child in Budget Decisions." *Science* 335(6064): 25–26. doi: 10.1126/science.335.6064.25

Malakoff, David. 2014, March 14. "The Future Is Flat in White House's 2015 Spending Request." *Science* 343(6176): 1186–87. doi: 10.1126/science.343.6176.1186

Chapter 12

Charles, Daniel. *Lords of the Harvest: Biotech, Big Money, and the Future of Food.* Cambridge, MA: Perseus, 2001.

Economic Research Service, USDA. *Agricultural Science Policy: Briefing Room.* Washington, DC: USDA. Retrieved January 2015, from http://www.ers.usda.gov/topics/farm-economy/agricultural-science-policy.aspx

Fuglie, Keith O., Nicole Ballenger, Kelly Day, Cassandra Klotz, Michael Ollinger, John Reilly, Utpal Vasavada, et al. "Agricultural Research and Development: Public and Private Investments under Alternative Markets and Institutions." Agricultural Economics Report no. 735. Washington, DC: USDA, Economic Research Service, 1996, May. Retrieved May 7, 2015, from http://www.ers.usda.gov/publications/aer-agricultural-economic-report/aer735.aspx

Fuglie, Keith, Paul Heisey, John King, Carl E. Pray, and David Schimmelpfennig. "The Contribution of Private Industry to Agricultural Innovation." *Science* 338 (2012): 1031–32.

Jordan, John Patrick, Gale A. Buchanan, Neville P. Clarke, and Kelly C. Jordan. *Leadership in Agriculture: Case Studies for a New Generation.* College Station: Texas A&M Press, 2013.

King, John, Andrew Toole, and Keith Fuglie. "The Complementary Roles of the Public and Private Sectors in U.S. Agricultural Research and Development."

Economic Brief no. 19. Washington, DC: USDA, Economic Research Service, 2012, September.

Chapter 13

Alston, J. M., J. M. Beddow, and P. G. Pardey. "Agricultural Research, Productivity and Food Commodity Prices." Giannini Foundation of Agricultural Economics, University of California. *Agricultural and Resource Economics Update* 12(2) (2008): 11–14.

Alston, J. M., C. Chan-Kang, M. C. Marra, P. G. Pardey, and T. J. Wyatt. "A Meta-Analysis of Rates of Return to Agricultural R&D: Ex Pede Herculem?" *Research Report* 113. Washington, DC: International Food Policy Research Institute, 2000.

Beintema, Nienke M., and Gert-Jan Stads. "Measuring Agricultural Research Investments: A Revised Global Picture." *Agricultural Science and Technology Indicators: Background Note.* Washington, DC: International Food Policy Research Institute, 2008, October.

Beintema, Nienke M., Gert-Jan Stads, Keith Fuglie, and Paul Heisey. "ASTI Global Assessment of Agricultural R&D Spending." Washington, DC: International Food Policy Research Institute, 2012, October. http://www.ers.usda.gov/data-products/agricultural-research-funding-in-the-public-and-private-sectors.aspx

Charles, Daniel. "A 'Forever' Seed Bank Takes Root in the Arctic." *Science* 312(5781) (2006, June 23): 1730–31.

Economic Research Service. *International Agricultural Productivity.* Washington, DC: USDA. http://www.ers.usda.gov/data-products/international-agricultural-productivity.aspx

Falcon, W., and R. Naylor. "Rethinking Food Security for the Twenty-First Century." *American Journal of Agricultural Economics* 87 (2005): 1113–27.

Pardey, P. G., and N. M. Beintema. *Slow Magic: Agricultural R&D a Century after Mendel.* Washington, DC: International Food Policy Research Institute, 2001.

———, S. Dehmer, and S. Wood. *Agricultural Research: A Growing Global Divide? IFPRI Food Policy Report.* Washington, DC: International Food Policy Research Institute, 2006.

Pardey, P. G., J. Roseboom, and N. M. Beintema. "Investments in African Agricultural Research." *World Development* 25 (1997, March): 409–23.

Sage, Rowan. "Supercharging the Rice Engine." *CSA News* 54(5) (2009): 4–6.

Chapter 14

Herdt, Robert W. "People, Institutions, and Technology: A Personal View of the Role of Foundations in International Agricultural Research and Development 1960–2010." *Food Policy* 37 (2012): 179–90.

Jordan, John Patrick, Gale A. Buchanan, Neville P. Clarke, and Kelly C. Jordan. *Leadership in Agriculture: Case Studies for a New Generation.* College Station: Texas A&M University Press, 2013.

Karina, Stephen J. *The University of Georgia College of Agriculture: An Administrative History, 1785–1985.* Athens: College of Agriculture, University of Georgia, 1985.

Martenson, Chris. *The Crash Course: The Unsustainable Future of Our Economy, Energy, and Environment.* Hoboken, NJ: Wiley, 2011.

Nye, Joseph S., Jr. "Shale Gas Is America's Geopolitical Trump Card." *Wall Street Journal* (2014, June 8).

von Caemmerer, Susanne, W. Paul Quick, and Robert T. Furbank. "The Development of C_4 Rice: Current Progress and Future Challenges." *Science* 336 (2012): 1671–72.

Index

CPSIA information can be obtained
at www.ICGtesting.com
Printed in the USA
LVOW04*2105300316
481486LV00001B/1/P